Natural Disaster and Women

Natural Disaster and Women

A Gendered Perspective

Edited by

Pradeep Kumar Parida

RAWAT PUBLICATIONS

Jaipur • New Delhi • Bangalore • Guwahati • Kolkata

ISBN 978-81-316-0946-0

Published by
Prem Rawat for **Rawat Publications**
Satyam Apts., Sector 3, Jawahar Nagar, Jaipur 302 004 (India)
Phone: 0141 265 1748 / 7006 Fax: 0141 265 1748
E-mail: info@rawatbooks.com
Website: www.rawatbooks.com

New Delhi Office
4858/24, Ansari Road, Daryaganj, New Delhi 110 002
Phone: 011-23263290

Also at *Bangalore, Guwahati* and *Kolkata*

Typeset by Rawat Computers, Jaipur
Printed at Chaman Enterprises, New Delhi

Contents

	Foreword	7
	Preface	9
	Acknowledgements	11
	List of Acronyms and Abbreviations	13
	List of Contributors	15

PART I
Introduction

1	Natural Disaster and Women: A Gendered Perspective *Pradeep Kumar Parida*	19

PART II
Gender Inequality, Susceptibility and Disaster

2	Gender and Disaster: An Overview of Differential Vulnerabilities *Sudha Sitharaman*	51
3	The Socio-cultural Aspects of Gendered Vulnerabilities to Disasters: Case Studies of Women-Headed Households, Post Mumbai Train Blasts, 2006 *Nalini Yadav*	63
4	Gender Difference in PTSD, Mental Health and Quality of Life: A Focus on Female Tsunami Survivors *Neethu P.S. and Surendra Kumar Sia*	83
5	Dealing with Gender Vulnerability through Mediated Communication in the Tamil Nadu Floods, 2015 *S. Mahalakshmi and Sunitha Kuppuswamy*	102

PART III
Women's Vulnerability and Responses to Disaster

6 Revisiting Odisha Super Cyclone of 1999: The Barriers and Coping Mechanisms of Disabled Women to Natural Disaster 127
Pradeep Kumar Parida

7 Impact of Nepal Earthquake of 2015 on Women: Challenges and Responses 153
Rishi Gupta

8 Women's Participation in Relief Work Utilizing Social Media Network 176
Shanmugaraj G. and Balasaravanan T.

9 Understanding Women's Experience of Trauma and Posttraumatic Growth: A Review Study 193
Alphonsa Jose K. and Surendra Kumar Sia

PART IV
Assessing Preparedness, Recovery and Mitigation

10 Gender Sensitivity in Disaster Risk Management: Lesson Learnt from the Chennai Floods of 2015 221
Suriya S.

11 Indigenous Knowledge System and Disaster Management among the Indigenous People of North-eastern Himalayas, India 238
Ajeet Jaiswal

12 Natural Calamities and Native Perceptions: Evolving Effective Rehabilitation Strategies 257
Jesurathnam Devarapalli

13 Disaster Diplomacy in South Asia: India's Humanitarian Assistance Programmes in Neighbourhood 272
Sibaram Badatya

PART V
Conclusion

14 Natural Disaster and Women: Towards a Sustainable Disaster Risk Reduction 307
Pradeep Kumar Parida

Index 323

Foreword

Natural disasters are occurring with amazing frequency. With the unique geo-climatic conditions, India is vulnerable to a large number of disasters due to changing demographics, socio-economic conditions, unplanned urbanization and environmental degradation. It is already recognised that disasters expose and highlight gendered imbalances in society which include both vulnerabilities and capacities, and social and economic imbalances arising from class, caste, disability and minority status. Women and children are most vulnerable during and after any disaster. Therefore, the issue of gender in connection with natural disasters is becoming a major concern for the policy makers all over the world. Although there is progress in gender issues related to Disaster Risk Reduction (DRR), the mainstream academicians, disaster managers and professionals often lack the knowledge required to address gender concern in DRR. It is in this context, the book *Natural Disaster and Women: A Gendered Perspective* is timely as there is a pattern of gender differentiation at all levels of the disaster process.

There is no doubt that academic institutions and scientific organizations have to play different roles in the direction of disaster risk reduction. As a part of the social science research, this book addresses several aspects of gender relations that cut across all segments of society. The book makes a major contribution to our understanding of gendered vulnerability and its impact on individuals and society. It underscores the diverse barriers that women face before and after disaster. Discussing these barriers and implementing in policy is definitely within our reach. Several chapters of this volume argue specific case studies which are unique in nature. Further, understanding the gender dimension of natural disaster is the need of the hour as we cannot afford to see the most vulnerable population face differential susceptibilities due to natural disasters. I, therefore, hope that this book will prove to be useful to the academicians, planners, researchers and policy makers in mainstreaming gender into DRR. In fact, this book offers new prospects to advance gender equality in the socio-economic development process. The implications for every action in this regard are in a broader context to move further towards sustainable development.

Vinod K. Sharma
Senior Professor, Disaster Management
Vice Chairman, Sikkim State Disaster Management Authority
Member, UNISDR, Asia Science, Technology and Academia Advisory Group

Preface

India and several other countries all over the world are vulnerable to a large variety of natural and technological disasters. As I write the preface to this volume, I feel deeply saddened to know how thousands of people are dying in India, particularly in Gujarat, West Bengal, Odisha, Bihar, Maharashtra, Rajasthan, Uttarakhand, Himachal Pradesh, Assam and Mumbai due to severe floods and other related hazard agents. While remembering the recent catastrophic events like the Odisha super cyclone 1999, the Indian Ocean tsunami 2004, the Uttarakhand floods 2013, the Jammu and Kashmir deluges 2014, the Nepal earthquake 2015 and the Chennai floods 2015, one can safely argue that the catastrophic impacts of disaster are going to increase with demographic expansion, technological change, economic development and related social and political dynamics. In the meanwhile, a plethora of evidence advocates that disaster aggravates existing development challenges, further exposing the vulnerability of the socially isolated and pushing those residing on the margins closer to the edge. If that is the case, the question arises, which groups are more vulnerable to the impacts of disasters?

This volume is an outcome of the national conference 'Natural Disaster and Women: A Gendered Perspective' held at the Department of Sociology, Pondicherry University, Puducherry, 6-7,

October 2016, where scholars assembled to discuss the social construction of gender vulnerability in disasters. Indeed, disaster research, which has concentrated previously on crisis response and recovery, is deficient without the immediate study of social processes and social structures connected with disasters, which includes knowledge about the vulnerability of people living in hazard-prone areas. Since gender difference is considered as an important variable, accordingly, the broad objective of this volume is to understand why and how disasters are a gender issue?

Among the lessons I have learned from the recent hazards is the significance of a sociological perspective in the scholarship of disasters, something which I once doubted. At present, the foremost problems in the Third World countries are not, in general, the absence of disaster management specialists; however, there is a problem with regard to understand the societally-induced vulnerability associated with places and populations. Within this broader context, the volume focuses on women's needs and concerns and demonstrates why and how gendered perspectives require to be integrated into the disaster risk reduction process. Our goal is to discuss how women are disproportionately vulnerable to various disasters. Knowing that disasters are socially constructed events that are part of the social process, authors in this volume have tried to establish the directions on how to integrate gender concerns in disaster risk reduction. Since more women are affected in disasters throughout the world, a comprehensive and detailed analysis would help in protecting one of the most vulnerable groups during disasters. *Natural Disaster and Women: A Gendered Perspective* is expected to offer a foundation for planning future social science disciplinary, multidisciplinary and interdisciplinary research as well as present a huge source for members in the policy making bodies to bring necessary improvements in policy measures.

Acknowledgements

As editor, I would like to acknowledge several people who made this volume a success. This book would not have been possible without the financial support from Indian Council for Social Science Research (ICSSR), New Delhi and Pondicherry University, Puducherry to organise a National conference on the same theme. To embark on this endeavour, I would like to acknowledge the efforts of those who presented papers in the National conference. Out of 47 presenters, 13 papers are selected for this book. My personal thanks go to all the authors who contributed papers.

There are many people to thank who helped me immensely: my department colleagues Prof. B. B. Mohanty, Prof. G. Ramathirtham, Prof. Sudha Sitharaman, Prof. Gulam Dasthagir, Prof. C. Aruna, Dr Imtirelna Longkumer, Dr Mansy M., Dr Aashita and Dr Thanuja Mummidi, who contributed to making the National conference a true intellectual feast. I am similarly grateful to Shri A. M. Giri, Prof. Surendra Kumar Sia, Dr Jeetendra Mishra, Dr Amaresh Samantaraya and Prof. Sampad Kumar Swain for their friendly support and Dr Binu Zachariah for his meticulous proofreading of my papers.

I am grateful to all victims of the Odisha super cyclone of 1999, particularly the women with disability for their support at the time of data collection.

My thanks also to our Department Guest Faculty, Dr R. Saravana Raja, Dr Hashim T, Dr Shafeeque K., Dr Dipti Ranjan Malik and Dr Kiran Kumar and my research scholars and students for helping me organise the national conference. I also acknowledge the support of Rawat Publications.

My deepest thanks go to Shri Manoj Parida, Prof. V. K. Sharma, Prof. D. Rajasekhar, Prof. Malabika Deo, Prof. Venkata Raghotham and Prof. Charu Malhotra who generously gave their time to support me to organise the national conference.

A special thanks to my son Pratinav and wife Latika for tolerating my considerable absence of mind and motivating me to work. I am really indebted to my parents, more particularly to my brother Atul Kumar Parida, a high school teacher, for his constant and utmost guidance in my academic endeavour.

This book is dedicated to Late Professor D. N. Dhanagare. It is a small repayment of the intellectual debt I owe.

September 2017 **Pradeep Kumar Parida**

List of Acronyms and Abbreviations

ADB	Asian Development Bank
BBC	British Broadcasting Corporation
CDP	Comprehensive District Plan
CEDAW	Convention on the Elimination of all Forms of Discrimination against Women
CSRC	Community Self Reliance Centre
DDMA	District Disaster Management Authority
DRR	Disaster Risk Reduction
DSM	Diagnostic and Statistical Manual
GDP	Gross Domestic Product
GII	Gender Inequality Index
HDI	Human Development Index
ICSSR	Indian Council for Social Science Research
ICNR	International Conference on Nepal's Reconstruction
IDNDR	International Decade for Natural Disaster Reduction

IFRC	International Federation of Red Cross and Red Crescent Societies
IMD	India Meteorological Department
IOM	The International Organization for Migration
LHD	Low Human Development
MMR	Maternal Mortality Ratio
MWCSW	Ministry of Women, Children and Social Welfare
NDMA	National Disaster Management Authority
NDRF	National Disaster Respond Force
NEC	National Executive Committee
NGO	Non-Governmental Organisations
NRA	National Reconstruction Authority
NSDRM	National Strategy for Disaster Risk Management
OVP	Organismic Valuing Process
PTG	Posttraumatic Growth
PTSD	Posttraumatic Stress Disorder
SDMA	State Disaster Management Authority
SHG	Self-Help Group
TASMAC	Tamil Nadu State Marketing Corporation
UNISDR	United Nations International Strategy for Disaster Reduction
UNDP	United Nations Development Programme
UNEP	United Nations Environment Programme
UNHD	United Nations Human Development
UNICEF	United Nations Children's Fund
USAID	United States Agency for International Development
VAW	Violence against Women
WEF	World Economic Forum
WHO	World Health Organization

List of Contributors

Pradeep Kumar Parida, Associate Professor, Department of Sociology, Pondicherry University, Puducherry.

Sudha Sitharaman, Professor, Department of Sociology, Pondicherry University, Puducherry.

Nalini Yadav, Research Scholar, Jamsetji Tata School of Disaster Studies, Tata Institute of Social Sciences, Mumbai.

Neethu P. S., Research Scholar, Department of Applied Psychology, Pondicherry University, Puducherry.

Surendra Kumar Sia, Professor, Department of Applied Psychology, Pondicherry University, Puducherry.

S. Mahalakshmi, Research Scholar, Department of Media Sciences, Anna University, Chennai.

Sunitha Kuppuswamy, Assistant Professor, Department of Media Sciences, Anna University, Chennai.

Rishi Gupta, Research Scholar, Center for South Asian Studies, Jawaharlal Nehru University (JNU), New Delhi.

Shanmugaraj G., Research Scholar, Department of Electronic Media and Mass Communication, Pondicherry University, Puducherry.

Balasaravanan T., Assistant Professor, Department of Electronic Media and Mass Communication, Pondicherry University, Puducherry.

Alphonsa Jose K., Research Scholar, Department of Applied Psychology, Pondicherry University, Puducherry.

Suriya S., Research Scholar, Madras Institute of Development Studies (MIDS), Chennai, Tamil Nadu.

Ajeet Jaiswal, Assistant Professor, Department of Anthropology, Pondicherry University, Puducherry.

Jesurathnam Devarapalli, Assistant Professor, Department of Anthropology, Pondicherry University, Puducherry.

Sibaram Badatya, Assistant Professor in Political Science, School of Law, Vellore Institute of Technology (VIT), Chennai.

PART I
Introduction

1

Natural Disaster and Women

A Gendered Perspective

Pradeep Kumar Parida

The tool of gender analysis is a powerful one for accurately diagnosing opportunities and constraints in any programme situation, and for identifying more effective strategies for delivering emergency assistance so that it supports long-term development for women and men, and girls and boys.

(Mary B. Anderson, Author)

If we are going to see real development in the world, then our best investment is women.

(Desmond Tutu, Nobel Peace Prize Winner)

Women have the power to shape our destinies in ways previous generations couldn't imagine.

(Hillary Clinton, Former U.S. Secretary of State)

Natural disaster is increasing at a great pace and intensity. A remarkable increase in the number of events and a geometric swell in the number of people impacted by disasters associated with natural hazards speak volumes of this burning issue (IFRC 2003).

Gone are the days when people thought natural disasters were due to the 'acts of God' or the 'wrath of nature'. Whereas a hazardous event does create damage in its path and for a brief period of time during and immediately following an event, disasters are largely the results of human agency. Since disasters are not caused exclusively by natural events, their impacts are not the same for every individual. A number of individuals and groups become victims, while others remain relatively safe and sound. At a deeper level, therefore, it is important to understand the structure and function of the society, particularly the socio-economic status, culture and household living arrangements. Consequently, researchers are more and more focused on social, economic, political and cultural factors that determine different levels of vulnerability to natural hazards among nations, societies and individuals (Blaikie et al. 1994; Enarson and Morrow 1998). Looking at the impact of disasters, it can be safely said that even within a single village there may be differential impact due to a variety of factors including socio-economic status, income, level of education, ethnicity, age or gender (Wisner et al. 2004). In a similar vein, Blaikie et al. (1994) say that disasters are the results of social, political, and economic environment, as well as the natural events that cause them. As a result, there is a paradigm shift in disaster research more particularly in sociological and anthropological scholarships that have broadened and supplemented traditional 'event' oriented disaster studies which focused too narrowly on disaster events, their immediate impacts and consequent responses. At present, disasters are viewed as events that transpire within specific historical contexts that are produced through interactions among political, economic, demographic and environmental processes (Blaikie et al. 1994, Bolin and Stanford 1998; Wisner et al.2004). A disaster event such as flood, cyclone, earthquake or tsunami may supply a trigger or catalyst; however, the roots of disaster lie in the societally induced vulnerability. The study of disasters therefore 'is not an isolated speciality, but is necessarily connected to ... issues of development, environmental sustainability, urban geography, political ecology and critical sociological theory' (Bolin and Stanford 1998: 3).

Since the 1970s, though 'gender' has been identified as a critical subject in development policy, there is lack of practice to contest the rhetoric. In 1975, the UN declared a 'Development Decade for Women', stating that humanitarian plans should 'give sufficient consideration to the interests of women' (Gardner and Lewis 1996: 66) and throw away their 'male bias' (Chant and Gutmann 2005: 240). The subject of women in development also surfaced with the essential initiative that women have varied needs, requirements and interests, and therefore development schemes should influence and support them in ways different from men. In spite of that, it is not surprising that the gender dimensions of disasters have chiefly been overlooked in planning, practice and research. This is despite the fact that the outcomes of disasters are very likely to be gendered (Enarson 1998). While the sex of a person signifies the biological differences between men and women, gender in particular refers to the dynamic social processes that are deeply rooted within society that lead to socially constructed gender roles, identity, gender norms and learned behaviour suitable for men and women (WHO 2015). In other words, gender outlines the relationships between women, men and others, their personality, life chances and opportunities (Enarson 2012). In fact, gender is now acknowledged as a 'social structure' to a greater extent (see, for example, Acker 1992; Lorber 1994; Rubin 1975). Like race and class, gender is a social relationship of difference and inequality that is pertinent because it manages and influences all dimensions of public life. 'Gender not only organizes institutions and identities, but also shapes language and knowledge and even the ways bodies are known, experienced, used, and regulated' (Alway et al. 1998: 178). Thus, comprehending gender as a structure that comprises the entire proportions of social life offers a way for examining and understanding the associates between diverse dimensions of social reality. Alway et al. (1998: 179) provided a clear understanding:

> At the same time, gender serves as a means of transmission. Here, the conceptualization of gender as a social relationship of difference and inequality that organizes and affects all dimensions of social life, rather than a role, is especially significant in that it allows struggle (between individuals or groups) to be as significant a mode of transmission as socialization or negotiation. Changes or struggles on any particular level of social reality reverberate through the other levels along the lines of connection established by gender.

To begin with, my contention is that gender has been a significant, yet relatively unrecognized factor in disaster risk reduction, more particularly in the developing countries. Although the issue of 'gender' is considered in the regular planning and developmental process, in natural disaster plans, the concern of gender is seen as 'luxury', more particularly in rapid-onset disasters (Myers 1994). Thus, Anderson (1994: 10) rightly depicts:

> In all societies, men and women experience different vulnerabilities and have different capacities as a result of their gendered roles. Sometimes these roles are very different and rigid; sometimes they are overlapping and fluid. In either case, the failure to identify gendered roles and to plan programmes with them consciously in mind has resulted in the inequitable delivery of disaster relief assistance, and inadequate attention to the potential long-term outcomes of short-term interventions. The tool of gender analysis is a powerful one for accurately diagnosing opportunities and constraints in any programme situation, and for identifying more effective strategies for delivering emergency assistance so that it supports long-term development for women and men, and girls and boys.

There is no doubt that gender strongly shapes an individual's response to disaster (Parida 2015a, 2015b). Women are more vulnerable to the effects of disasters for a number of reasons. Literature reveals that there are biological, social and economic conditions and processes that force women to become more susceptible to natural disasters (Enarson 2000; Parida 2010). Owing to the mutual outcome of these conditions and processes, women possess different gender identities to men which show the way to their roles and responsibilities in society, and these identities, roles and responsibilities cause gender-based differences in a disaster society (Ariyabandu 2009; United Nations 2009). Especially, women's role as mothers, primary caregivers of the elderly, disabled and children executes a distinctively important role in deciding their disaster vulnerabilities (Enarson 2000; Enarson and Fordham 2001). The poor and deprived women are more likely to be susceptible to the outcomes of disasters. It is observed that 70 per cent of the people in the developing world living below the doorstep of poverty are women (Denton 2002). Taking a World Bank study, Johnsson-Latham (2007) points out that gender differences are greatest among the poorest families. Moreover, Oldrup and Breengaard (2009: 47) point out, 'In developing countries, women's needs are often not taken into consideration, and their participation in the climate change processes and debates is not sufficient at the national level'. Women and children are also 14 times more likely to die than men during disasters (e.g., Brody et al., 2008).

While drawing on specific conclusions from various literatures, Enarson and Fordham (2001: 134) explain the following processes which can increase women's vulnerability during disasters:

1. *Biological*

- Pregnancy and lactation increase vulnerability because of constrained mobility and the greater need for food and water; and
- Longer lifespans can lead to increased poverty in the elderly female population, contributing to the feminization of poverty.

2. *Economic*

- Reproductive work of bearing and caring for children and providing other essential family support, on which much male work in the public sphere depends, does not figure in economic (monetary) evaluations.
- Women's time inputs are not recognized and can be ignored when post-disaster relief and rehabilitation resources are being distributed.
- Women's external work opportunities lie disproportionately in part-time, temporary and low-status occupations which place them at greater risk of poverty.
- Lack of access to credit reduces women's ability to recover from disaster.
- Structural adjustment programmes exacerbate the negative impacts of increasing flexibilization of work and decreasing social programmes, contributing to the feminization of poverty, and making poor women targets for population control and violence.
- Priority for work opportunities may be given to men.

3. *Social*

- Widespread gender inequality in access to educational opportunities; and
- Illiteracy (as estimated by UNESCO for 2000, 14.7% for men and 26.4% for women).

4. *Political*

- Lack of universal suffrage; and
- Limited access to, and occupation of, decision-making power structures.

5. *Environmental*

- The domestic environment, especially in remote rural areas, can be particularly vulnerable in disasters but attention is generally focused on how public/commercial realms are impacted.

- The domestic environment is also a work environment for women earning income at home, who are doubly impacted by housing loss.
- Many women support families through homestead gardens and local agricultural production but these activities are not featured in assessments of economic production values nor are likely to be prioritized in disaster recovery programmes (Enarson and Fordham 2001: 134).

As I have noted elsewhere, gender and other constitutive components of social inequality have obtained limited attention in disaster scholarship, nevertheless, as a premier researcher in this field, Enarson (2000: 5) has systematically elucidated vast asymmetries and gender differences with regard to natural disasters:

- Post-disaster mortality, injury and illness rates which are often (but not universally) higher for girls and women;
- Economic losses which disproportionately impact economically insecure women (e.g., agricultural losses of women farmers, the destruction of women's home-based businesses, limited access to post-disaster economic aid);
- Work load changes which suggest that disasters increase women's responsibilities in the domestic sphere, paid workplace and community through the disaster cycle of preparation, relief, reconstruction and mitigation;
- Post-disaster stress symptoms which are often (but not universally) reported more frequently by women;
- Increased rates of sexual and domestic violence against girls and women in disaster contexts.

The pioneering works of Sen (1981, 1988, 1990, 2000) on the gendered nature of famine in developing countries present insights into how gender influences the distribution of resources in disasters. At a time when food is limited, women and female children take a lesser share of available food than men and boys. Households headed by women are at an extreme disadvantage following disasters because of their pre-existing social conditions. In this regard, Wiest (1998: 65) observes:

> Prior to an emergency or disaster, women in many societies are already subjected to a disproportionate amount of stress. The social and economic vulnerability of women rooted in unequal entitlements, unrecognized and unremunerated labour and

shortage of resources is often accentuated by familial conflict in the face of the ongoing need to care for dependent children and provide for family. In as much as the continuing integrity of the household and family disproportionately depends upon women, household and family disruption or separation associated with disaster weakens the coping capacity of the individual members.

Thus, women are differentially vulnerable in numerous circumstances. Generalizing about it tells us about the arrangement of social relations of power in particular contexts.

However, one caveat deserves to be mentioned here. Although abundant literature reveals women's differential vulnerability and disaster risk, it would be a mistake to characterize women as 'victims' and as wholly unable to cope with hazards and disasters. While lacking access to the broad range of supports, the at-risk women have their own internal supports and networks on which they can rely when disaster strikes. Thus, vulnerable and marginalized women are also resilient. The planning and developmental challenge is to engage the indigenous coping mechanism of communities at risk at all stages of the disaster cycle before, during and after disasters. In fact, the intent behind this introductory chapter is to create awareness about the state and other disaster reduction mechanisms to offer fitting support that can transform at-risk groups from possible disaster victims to active agents in the disaster risk reduction.

In recent years, extensive literature shows that women as a group are prone to respond experience and get affected by disasters differently (Enarson and Morrow 1998; Parida 2010, 2014). However, at the same time, it is also important to discuss how women are not seen as a homogenous group. Multiple factors, for example, class, caste, kinship, age, nationality and sociocultural groups are important variables. Analysing these differences is as vital as analysing the differences between women and men (Kelkar and Nathan 1991). Understanding the specific context in which a woman experiences a disaster within given set of social, economic, political and even religious circumstances is critical to know the broader subject of the gender dimension of disaster. In other words, more than recognizing a male–female differentiation, a deeper and in-depth understanding of how gender relates to the complex interplay of power, resources, privilege and stratification would enhance our ideas on this issue. Accordingly, the following women are considered to be highly vulnerable and most hard-hit before, during and after disasters strike (Enarson 2000).

- destitute, low income and economically insecure women; women who are contingent workers or unemployed; home workers and others in the informal sector; small-scale farmers; women in caregiving jobs and professions;
- women in subordinated racial/ethnic/cultural groups; recent immigrants and undocumented women; women migrant workers;
- women heading households, those in large complex households, and women caring for any dependents;
- frail senior women, undernourished women and those with chronic health problems or disabling physical and mental conditions, and women whose mobility is restricted due to pregnancy or childbearing;
- widows and single women; socially isolated women; rural women;
- women subject to domestic and sexual violence, and those insecurely housed in shelters; and
- functionally illiterate women; women not fluent in majority languages.

Major Approaches in Gendered-based Disaster Studies

Let's begin by examining briefly the main theoretical formations that have directed gender disaster investigation to date. Although no single theoretical framework guides disaster research on gender, majority of the researchers employ the social vulnerability approach (Bolin et al. 1998; Hewitt, 1983, 1997; Sen 1981; Wisner et al. 2004). Vulnerability science originated with a plethora of research undertaken both in the West and in the developing world (See Blaikie et al. 1994, Cutter 2003; Hewitt 1983; Sen 1981). Vulnerability analyses have argued that natural disasters are outlined by the pre-existing social, political, environmental and economic conditions and thus should not be regarded as 'natural' happenings (Cannon 1994). In this regard as Quarantelli (1990: 18) says, 'there can never be a natural disaster; at most there is a conjuncture of certain physical happenings and certain social happenings'. Wisner et al. (2004: 11), in the most systematic statement of hazard vulnerability research to date, define vulnerability as 'the characteristics of a person or group and their situation that influence their capacity to anticipate, cope with, resist, and recover from the impact of a natural hazard'.

UNISDR (International Strategy for Disaster Reduction 2004) defines vulnerability as 'the conditions determined by physical, social, economic and environmental factors or processes which increase the susceptibility of a community to the impact of hazards'. Based on the above explanation, Ginige et al. (2009: 26) illustrates the different dimensions of vulnerabilities as follows:

- physical vulnerability refers to susceptibilities of the built environment and may be described as 'exposure';
- social factors of vulnerability include levels of literacy and education, health infrastructure, the existence of peace and security, access to basic human rights, systems of good governance, social equity, traditional values, customs and ideological beliefs and overall collective organizational systems;
- economic vulnerability characterizes people less privileged in class or caste, ethnic minorities, the very young and old, the disadvantaged and often women who are primarily responsible for providing essential shelter and basic needs; and
- environmental vulnerability refers to the extent of natural resource degradation.

On the other hand, adopting from McEntire (2001), Ginige et al. (2009: 27) categorize the variables, which interact to produce increased vulnerabilities under physical, social, cultural, political, economic and technological titles as given in the following:

1. *Physical*

- the proximity of people and property to triggering agents;
- improper construction of buildings;
- inadequate foresight relating to the infrastructure; and
- degradation of the environment.

2. *Social*

- limited education (including insufficient knowledge about disasters);
- inadequate routine and emergency health care;
- massive and unplanned migration to urban areas; and
- Marginalization of specific groups and individuals.

3. *Cultural*

- public apathy towards disaster;
- defiance of safety precautions and regulations;

- loss of traditional coping measures; and
- dependency and an absence of personal responsibility.

4. *Political*

- minimal support for disaster programmes amongst elected officials;
- inability to enforce or encourage steps for mitigation;
- over-centralization of decision making; and
- isolated or weak disaster-related institutions.

5. *Economic*

- growing divergence in the distribution of wealth;
- the pursuit of profit with little regard for consequences;
- failure to purchase insurance; and
- sparse resources for disaster prevention, planning and management.

6. *Technological*

- lack of structural mitigation devices;
- over-reliance upon or ineffective warning systems;
- carelessness in industrial production; and
- lack of foresight regarding computer equipment/programmes.

The sociopolitical ecology perspective, most evidently employed by Peacock et al. (1997) in their book on 'Hurricane Andrew', argues about the broad ecological and political approach. This way of thinking about disasters is concerned with conflict, competition and inequality, rejecting the notion that a community is a single, autonomous social system (Peacock and Ragsdale 1997). This perspective 'was consistent with increased in political economy and critical perspectives, including the analysis of minority, gender, and inequality issues at all phases of disaster research' (Peacock et al. 1997: 21).

Examining gender relationships in particular environmental setup, the feminist political ecology perspective describes women's environmental knowledge and the association between gender inequalities, environmental degradation and disaster vulnerability (Rocheleau et al. 1996). While depending on natural resource based livelihoods, women exhibit tremendous responsibilities to the dependants in the household. Thus, they are quite aware of any natural disaster that put their families and homes at risks (Steady 1993). Indeed, from an eco-feminism point of view, feminist political ecologists have discussed gendered environmental knowledge and female leadership in grass-roots movements against the destruction of natural resources.

Feminist theory also played a crucial role in understanding the issue of gender before, during and after disasters strike. Enarson and Phillips (2008) describe that the sociology of disasters and feminist theories create a closer association because they use similar concepts such as social power, privilege, domination, vulnerability, empowerment, political economy and social change and likewise employ global, interdisciplinary and practice-oriented inquiry. While disaster researchers put forward that disasters break off 'the social system', feminist theory asks the question 'whose social system?' (Enarson et al. 2007).

Liberal feminist theorists have made important and extensive contributions to understand the gendered nature of disaster. Their entire focus is to recognize the gender differences, which according to them are socially created with distinctive human rights (Enarson et al. 2007). They relate social inequality in terms of access to education and achievement, cultural devaluation of women as well as the gendered division of labour, gender violence and limitations on reproductive choice (Lorber 1998). Scholars have generally attempted to examine 'how gender stereotypes affect disaster services and emergency operations; the careers of women in the field; and gender bias in the design, funding, implementation, monitoring, and evaluation of emergency shelters, water and sanitation, health care, and other post-disaster initiatives' (Enarson et al. 2007: 132).

The perspectives of ecofeminism and feminist environmentalism have dominated developmental discourse reflecting women's potential to regenerate resources within the culture of male domination and exploitation. Shiva (1988) says that the systematic turmoil of the 'feminine principle' makes women to face marginalization and displacement. Being closer to nature though women have potential to conserve the resources, their subordinate position makes them passive in the production process. Similarly, Agarwal (1992) argues that gender, class, caste and ethnicity interact with nature in creating environmental degradation and vulnerability among women.

In addition to the above, the principle of gender and development theory has been an indispensable lens for scholars studying a diverse range of gendered disaster vulnerability (e.g., Fernando and Fernando, 1997; Tinker 1990). Indeed, this theory posits the importance of 'patriarchy' and the historical dynamics of global capitalism and colonialism. Historically, women have been at the receiving end due to the dominant global culture, for instance, how free trade policies challenge and destabilize local markets and amplify pressure on men to migrate for earnings, leaving more women and children impecunious in untenable rural environments or displacing to hazardous urban settlements. Women's socio-economic and psychological background is a point of investigation with their coping strategies in risky environments. More particularly, the heightened interaction between marginalized groups such as female migrants and refugees, women agriculturalists, street vendors, home-based workers, single mothers, widows, impoverished and low-caste women and the nature of extreme susceptibilities in different contexts is always a matter of further research. Scholars have also theorized women's self-protective action in risky environments based on their reproductive, productive and community work. From this perspective, women are not only considered as the 'victims' but their basic fundamental human rights are questionable in disasters (Enarson and Fordham 2004).

Gender, Dependency and Natural Disaster

Importantly, effective control over and access to natural resources such as land, water and forests are important indicators of gender positions (Dankelman 2002). The use and management of these resources as well as decision-making at micro and macro-levels are gender differentiated. It may be said here that the association between women and their environments are not always constructive. For women, arable land is the essential source for meeting food needs and livelihoods. In most of the developing countries, women are at the hub of subsistence food production. Nevertheless, the UN statistics show that women own no more than 1 per cent of the world's land, and yet while they have access to it for cultivation, their tenure is often costly and uncertain. They are denied access to credit, training, and other supports to production without possession of land or safe access to it (Davidson 1993). Women are also considered as the 'invisible' water managers, accountable for supplying the water requirement of the family unit, household animals and occasionally cultivation. Voluminous studies have confirmed how their responsibility in carrying water over long distances is important for the health, economy and social development of local communities.

Nonetheless, they are often debarred from the planning, implementation and preservation of water supplies (Davidson 1993). Forest is an integral part of the lives of rural poor women. Trees are essential not only in safeguarding watersheds, controlling water flows, preserving soil fertility and air quality but also benefitting women's income-earning activities in terms of food, fodder, fuel, building materials and medicines. Rising rates of deforestation have deeper socio-economic and environmental consequences on the lives of poor women. The Centre for Science and Environment (CSE), India's Environment Report in 1985 says:

> Probably no other group is more affected by environmental destruction than poor village women. Every dawn brings with it a long march in search of fuel, fodder and water ... As ecological conditions worsen, the long march becomes even longer and more tiresome. Caught between poverty and environmental destruction, poor rural women in India could well be reaching the limits of physical endurance. (CSE, 1985: 172)

The Chipko Movement is an example of how women have fought desperately to protect forests for their daily survival. Thus, Shiva argues that 'economy and ecology are not opposed but converge in the survival economies of the third world poor however, in the context of market-oriented development they have been rendered contradictory' (Shiva 1991). Similarly, the condition of urban poor women is even worse than that of their rural sisters due to several factors such as hazardous health, environmental degradation and resource depletion. Most of them live in the squatters that are often built on mud and thatched. They do not have basic services such as regular water supply, sewerage, garbage removal or electricity. As a result, they are more prone to landslide, flooding, cyclone and tsunami.

All in all, the woman as a 'victim' argument depends on the differences between women and men's knowledge, responsibilities and control over resources. A UNRISD/UNFPA study on the 'Social Impact of Environmental Degradation at the Community Level' discovers three hypotheses:

1. The local system of resource use are engendered and the male and female resource related livelihood systems are differentially affected.

2. The burden falls disproportionately on women because of their nurturing function. It is much easier for men than women to generate income in other spheres of economic activities.
3. Women's work obligations are not matched by a corresponding control over resources which would enable them to take the best trade-off decisions in circumstances of environmental decline. This reflects women's disadvantages in other dimensions of livelihood (market-related) in which environmental resource entitlements are embedded (UNRISD 1992).

From the standpoint of women's daily life, it is argued that 'women are trapped into a material dependence upon a man by lack of access to a decent wage, the absence of services to make full-time waged work and childcare compatible, the ideological and social pressures to marry, and the almost complete absence of any alternative to the family as a way of life, a learned sense of inferiority, and the personal effects of isolation and overwork' (Swaminathan 1987: 37, cited in Bland et al. 1978). Mothering has been a time-consuming activity for women; therefore, preparing themselves at the time of disaster has become a Herculean task. In discussing women's paid and unpaid labour, Glazer (1980) notes:

> In the nineteenth century, mother's work appeared at first to be the relatively simple tasks of seeing that children learned certain moral standards, were fed, were kept clean and slept regular hours ... This century mothers have gradually come to be responsible for building personality too. They must develop capabilities in their children during early childhood that will be compatible with demands of the school, the workplace and political life ... Mothers are responsible for the happiness of their children as well as their general fitness for life in a capitalist society. (pp. 260–261).

Similarly, stating about women's most deplorable conditions and vulnerabilities in causing disasters, Anderson (1994: 8) says:

> In general, around the world, women are poorer than men. Their poverty arises from the roles they are assigned and the limits placed by societies on their access to and control of resources. Women are disproportionately employed in unpaid, underpaid and non-formal sectors of economies. Inheritance laws and traditions, marriage arrangements, banking systems and social patterns that reinforce women's dependence on fathers, husbands and sons all contribute both to their unfavourable access to resources and their lack of

power to change things. The health dangers that result from multiple births can contribute to interrupted work and low productivity. Traditional expectations and home-based responsibilities that limit women's mobility also limit their opportunities for political involvement, education, access to information, markets, and a myriad of other resources, the lack of which reinforces the cycle of their vulnerability.

Gender Differentiation during the Process of Disaster

Based on Fothergill's (1996) nine-stage typology: (1) exposure to risk; (2) risk perception; (3) preparedness behaviour; (4) warning communication and response; (5) physical impacts; (6) psychological impacts; (7) emergency response; (8) recovery and (9) reconstruction, I have examined various social issues with regard to gender roles and relations during the process of disaster (adapted from Fothergill 1996). By taking stock of what is acknowledged in this area, my goal here is to illustrate gender differentiation during disaster preparedness, impact and recovery and reconstruction period. This permits us to identify how risk is distributed among various members of society during the process of disaster.

1. Exposure to Risk

It has already been discussed elsewhere that gender strongly influences vulnerability to disasters and risks. Women's susceptibility to bear the brunt of disasters is enhanced by socially determined differences in roles and responsibilities pertaining to their inequalities in access to resources and decision-making power. Particularly, impoverished women face greater exposure and risk because they lack status, power and resources. There is a sense of isolation that prevails and disconnects them from the larger system that ultimately augments risk exposure and susceptibility. Studies have shown adverse reproductive outcomes following disasters, including early pregnancy loss, premature delivery, stillbirths, delivery-related complications and infertility (WHO 2002).

It is noteworthy to state that women who struggled in their daily lives and lived in miserable conditions are doubly exposed to risks. In line with this, extant literature shows how poor women with financial insecurity and low income are time and again exposed to the outcomes of disasters leaving them with limited social choice, more practical restrictions and less recovery resources (Bradshaw 2001; Kafi 1992). Generally, low income women are predisposed to occupy housing that exposes them to destruction, dwelling in poorly maintained public housing, manufactured homes, shelters and rental properties more often than low income men (Enarson 1999). The Odisha super cyclone of 1999 reveals how women who resided in flimsy housing near the seashore faced the disastrous effects of the fury (Parida 2002, 2009). Subsequent to Hurricane Andrew, deprived women of colour were studied to be those who were most in need of housing even many years after the disaster (Morrow and Enarson 1996). Practically, poor and deprived women who have limited economic resources such as land, crop failure insurance, access to labour and tools are required to reconstruct their lives and homes subsequent to a disaster (Ollenburger and Tobin 1998). Moreover, within the poor and deprived women, the elderly, the physically challenged and the women-headed household find it extremely difficult to face the consequences of disaster due to their uncompensated tension because of physical weakness and societal processes (Parida 2016).

A burgeoning disaster literature has found how people are marginalized by race, ethnicity and caste divisions because of their tradition, culture, language, experience, stereotypes, discrimination, segregation and social isolation (Fothergill et al. 1999, Parida 2008). The intersection of poverty and race/ethnicity/caste combination mostly affects women who face the pre-existing housing-related difficulties coupled with discrimination in relief systems following a disaster. Numerous cross national studies corroborate this explanation. For example, Enarson and Fordham (2001) in their study of the 1997 Grand Forks flood in North Dakota found that disaster relief was taken away from migrant workers, affecting mainly Hispanic single mothers. Echoing these themes, Enarson and Morrow (1998) say that women who are already marginalized by racial/ethnic discrimination or economic segregation are less likely to take an active part in long-term disaster recovery process than more privileged women in dominant racial groups. From India and Sri Lanka, Parida (2008, 2015a) reveals how lower caste women and girls are vulnerable to sexual abuse and exploitation which compelled them to initiate migration.

2. Risk Perception

Both men and women perceive threats differently. In contrast to men, who see risks as part of life, women are hesitant to take risks. Research reveals that an individual's perception of risks is influenced by social factors, cultural biases and the networking mechanisms (Fothergill 1996). Generally, women are more serious to any external threats than their counterparts. It reflects their caring attribute to family and kinship. However, due to their vulnerable conditions, women's capacities to avoid or cope with and recover from disaster results from a complex interplay of political, economic, social and ideological practices present at a given locale and vary by a given hazard and by specific household characteristics (Blaikie et al. 1994).

3. Preparedness Behaviour

It is important to recognize that major disasters need preparations at different levels with proper coordination, communication and cooperation between them. Proposing an effective risk management framework, Britton (2001) mentioned the following elements: 'establishing the context'; 'identifying the risks'; 'analysing risks'; 'evaluating risks'; 'treating risks'; 'monitoring and reviewing' and 'communicating and consulting'. Unfortunately, women as category find it extremely difficult to prepare for any intended disaster. Generally, a disaster mitigation preparedness strategy is coordinated and consistent with a set of goals, policies and tools for reducing or minimizing losses from hazards and resulting disasters. But the fact of the matter is that neither have women the power and ability to coordinate disaster preparedness nor are they appointed in the governmental policy level as a planning manager for strategizing future disasters.

More generally, multiple factors are discussed elsewhere related to the pre-existing social conditions of women. In a way, women's preparation to impending disaster is dependent on the dynamics of community empowerment and their differential adaptive capacities.

4. Warning Communication and Response

Warning communication is an important mechanism for safe evacuation process. Effective warning and subsequent actions frequently save lives at the time of critical emergency. Research shows that men and women receive warning communication differently. Because of women's social networks and roles, they receive risk communication effectively and try to initiate protective actions such as evacuation (Fothergill 1996). More women interpret the disaster warnings as valid (Drabek 1969). This behaviour corroborates women's heightened perception of risks. However, evacuating a place depends upon a complicated interplay of economic, social and ideological practices. It is a complex process involving the participation of numerous organizations and actors working in coordination towards a common goal. As in any complex coordination structure, it is imperative for all pertinent parties to recognize their roles and tasks within the broader response framework in order to guarantee well-organized and successful evacuations. Although women internalize the warning more often than their counterparts, the decision of evacuation is taken mostly by the male members (Parida 2010). Moreover, if there is evacuation, the burden of children, the elderly and the physically challenged lies on the shoulders of women. Even after reaching the new place, it is the responsibility of women to pay attention to all family members by means of feeding them in a difficult and competitive environment.

5. Physical Impacts

Voluminous literature shows that women's caregiving responsibilities place them more at risk of injury and death as they always attempt to save their children and elderly members of the family. Given that women are physically more impacted following disasters, there is no doubt that gender differences in fatalities can be basically attributed to the daily ways of life that situate women at higher risk (Ikeda 1995; Rivers 1982). The 2004 Asian earthquake and tsunami is a blatant example where more than 400,000 people were killed in 12 countries. In most regions and districts, up to 80 per cent of deaths were women and children (Oxfam 2005, cited in Amaratunga and O'Sullivan 2006). The 2004 tsunami illustrates that the morbidity and death rates following the disaster are disproportionately high for women, children and the elderly (Parida 2015a). Particularly girls and women were vulnerable because of social and cultural factors, including their roles as family caregivers. In Matara, southern Sri Lanka, countless women were standing at the bus stands and morning market adjoining to the sea. Since women and girls are rarely taught to swim, unlike men and boys, women found difficulty running away from danger because of their lengthy, wrap-around clothing. In Sri Lanka, several women drowned trying to save their children; others felt traumatized since they could not save all of their children (Amaratunga and O'Sullivan 2006). Similarly, in the Odisha super cyclone of 1999, more women died as they were living in flimsy houses close to the sea (Parida 2004, 2008).

As in the case of significant physical impacts and high death rates, however, research has indicated three main causes for gender differences. Firstly, biological and physiological differences between men and women from time to time disadvantage women during the process of disasters. Secondly, social norms and role behaviour construct the behaviour of women, which enhances their susceptibility before, during and after the disaster. Finally, disasters more often generate scarcity of resources as well as breakdown of social order. As a result, the competition between individuals becomes severe and existing forms of gender discrimination and sexual violence become aggravated. With the exception of biological and physiological factors, the higher susceptibility of women is socially constructed and is due to differences in the socio-economic status of men and women (Neumayer and Plumper 2007).

6. Psychological Impacts

The impact of disaster generally produces tremendous psychological effects, such as anxiety, depression, avoidance and intrusion. Within the mental health literature, Peuler (1988: 259) says 'personality traits culturally reinforced by sex roles create observable differences in the impact of the disaster on men and women ... Some typical traits observed in women group members were the tendencies to put others first, or to feel overly responsible for the welfare of others'. The point is that while men and women have primarily diverse 'typical' traits, presumably psychobiological in origin, these are strengthened by explicit sex roles that people simply learn and perform. So men and women have separate universal constructs based on the historical and ongoing cultural practices that have structured the social and psychological impacts of disaster. Thus, to understand the gendered nature of disasters' mental health impact, we need to recognize the wider picture of major variations. Let's take the example of health of women in different social groups. It is understood that women's physical health and mental health are interrelated to each other. Usually the negative and disastrous effect of a catastrophe is not the only factor for expressing more mental health problems; nonetheless, the day-to-day pre-disaster health issue is extremely important to know the overall mental and physical well-being. Literature points out that several women in Third World countries try to cope with the broader health consequences of poverty in the forms of communicable diseases and undernutrition (Doyal 2001). Although women face the same endemic diseases as men, both biological and social issues amplify their exposure before, during and after disaster. While malaria, hepatitis and leprosy, for example, can be in particular risky during pregnancy, women's responsibility for domestic tasks increases their chance of contracting water-borne diseases during flood, cyclone and other disasters.

The case of undernutrition in girls and women is important with regard to mental health issues following disasters. It is estimated that around 44 per cent of all women in Third World countries are anaemic compared with about 12 per cent in developed countries (WHO 1992). In India the number is as high as 88 per cent (World Bank 1993). Moreover, data from many Third World countries advocates that women obtain less health treatment of all kinds than men, in spite of their greater need to avoid casualty at the time of disaster. The danger here is that most rural women are deprived of getting access to modern services even for obstetric care (United Nations 1991). This speaks the volume about the general health condition of women in their day-to-day life which has a deeper connotation in the augmentation of mental health issues. Sometimes we dwell on the positives of female life expectancy which continues to increase in most Third World countries without talking about the pain of the elderly women.

The departure point is that although women show 'flexibility' and 'adaptability' due to their traditional role of home maker and caregiving to others (Clason 1983), they express greater amount of stress than men due to their pre-disaster health and socio-economic conditions. Disaster works like a double-edged sword for them amplifying severe pre-disaster mental health problems such as stress, depression and post-traumatic stress disorder (PTSD). Since women face the burdens of care giving on a daily basis, any casualties or destruction (real or anticipated loss of life, physical injury, damage or economic loss) at the family level increases their rate of depression. However, studies also depict how men experience acute anxiety when they are separated from their family (Fritz and Marks 1954), though, this may be for a brief period of time, lesser than their counterpart. Thus, while both men and women experience anxiety, women often suffer from acute and greater mental stress that persists for many years.

7. Emergency Response

Our knowledge of emergency response is all about responding to a situation which demands coordination and quick decision at various levels. It is a complex mechanism which is rooted in several developmental cycles and cultural constructions. Thus far, developmental cycle transformations have disproportionate effects based on the 'size and composition of domestic groups, the nature of inheritance, and resource availability and control'. Cultural rules including patriarchal ideology and a strong patriarchal bias in the control of resources are very much evident in developing countries. The careful argument is that:

> Peoples' responses to emergencies are grounded in an ongoing domestic social structure that simultaneously provides the context, form, and meaning for response—and is also a critical source of disaster vulnerability. The avenues through which people gain access to resources, social status, and even psychological well-being have an important bearing on their perceptions of hazard, disaster mitigation, and recovery. Disasters can destroy or fragment even well-endowed households, but if the ability to cope with day-to-day events is already stressed, households are particularly vulnerable. (Wiest 1998: 63–64)

Thus, before we understand women's disaster emergency response, it is inevitable to know about their pre-disaster 'cultural meaning in social relationships and structures of inequality that condition people differently' (Wiest 1998: 64). The fact of the matter is that women and their dependent children are inexplicably hindered by poverty and impoverishment. Whether it is agrarian or industrialized societies, the fate of women is one of formal dependency and, therefore, 'structural subordination'(Agarwal 1990). Disasters usually exacerbate the already existing severe inequalities in society by facilitating the concentration of resource control in the hands of a small percentage of population. So, women already underprivileged in terms of resource access are frequently at great risk.

As expected, family plays a crucial role in both spontaneous and organized response to disasters particularly in the developing world (Parida, 2002, 2015b). However, it is often endangered with the impact of disasters as a result of familial organization in terms of patriarchal ideology and male as 'bread winner' and social security comes under serious setback due to access to resources. Subsequent to this, rising landlessness and a weakly developed market leave the majority of the rural population without such security, heightening the oppression of women (Wiest 1998).

However, it is also argued that women play a positive role in crisis service organizations, due to their greater compassion, sensitivity to victims and family management abilities. In this context, Phillips (1990: 90) reveals that women are 'sensitive to victims' and workers' needs, such as knowing when to say a word of encouragement or support;' and 'female presence introduced more civility and an emotionally supportive environment'.

8. Recovery

The heightened interaction between women's multifaceted roles concerning productive, reproductive, community labour and disaster shows how the gendered division of labour at home, mainly the caregiving roles increase women's pre-disaster susceptibility and as a result accelerates burdens during recovery. On the other hand, various women's groups are often keenly engaged in the emergency period, delivering food and other supplies. These groups are basically the pre-existing women's community groups that render services during the emergency and recovery period. Although immediate disaster emergency witnesses a gender-integrated approach (Fothergill 2004), the recovery period further demonstrates gender division in the decision-making and planning process (Noel 1998). Several studies reveal women's underrepresentation in the emergency management system. For example, while Khondker (1996) found that women were debarred from disaster response decision-making activities in Bangladesh, Begum (1993) as well as Parida (2016) narrated how fewer relief supplies were provided to women in need. Thus, there is a clear evidence of gender bias in disaster risk reduction decision-making process.

9. Reconstruction

Generally reconstruction is 'characterised by rebuilding infrastructure, restoring community services, securing loans and long term assistance, and attempting to return to a somewhat normal life and routine' (Fothergill 1996: 47). Although the state with institutionalized welfare mechanism tries to provide disaster assistance (Wiest 1998), nonetheless, from a gender perspective, it is clear that poor and deprived women find it most difficult to avail assistance and return to their normal life. Since poor and deprived women have less insurance and less savings, there is less possibility of long-term recovery (Bolin and Bolton 1986). Various studies show how the state mechanisms have failed to identify the gender concern and women's short-term needs and long-term interests, although social class, race and ethnicity were found to be powerful determinants of aid following Hurricane Andrew (Enarson and Morrow 1997; Yelvington 1997). Due to technical and bureaucratic procedures, poor women more often did not get the aid to which they were 'entitled' because relief assistance was rooted in a single 'head of household' model that tended to privilege men (Morrow and Enarson 1996). In a similar vein, Childers (1999) discovered that low income elderly women were disproportionately supplied economic assistance, although they needed it the most and were considered at high risk.

While the above list is not complete, it presents a glance of the issues that must be addressed in gendered disaster research and operational practices. However, further research is essential both at the local, regional and national and at international levels to make it more comprehensive.

Conclusion

Women are vital in all society including disaster-prone societies. As argued above, most of the women's susceptibility to the brunt of disasters is amplified by socially determined disparities in roles and responsibilities. Since aspects of gender intersect all parts of society and have implications for every action, a comprehensive understanding of this issue will lead towards meaningful disaster mitigation and sustainable development. As Bolin et al. (1998: 43) rightly stated, 'Knowing to what extent women, and other historically marginalised groups have particular vulnerabilities to various kinds of natural and technological hazards facilitates the development of appropriate assistance programmes, ones that don't exacerbate social inequalities or privilege men over women or households over individuals'.

Organization of this Volume

The book is divided into five sections. The introductory chapter falls in Part I, which sets the theme of the book. After briefly putting various approaches, it elucidates gender, inequality and vulnerability issues in natural disaster. It also throws light on how natural disaster is a gender issue by examining more specifics of social issues in the disaster process. In Part II, contributors highlight gender inequality, susceptibility vis-à-vis disaster. In Part III, women's vulnerability and responses to disaster is explained by taking a range of case studies from various developing countries including India. In Part IV, gender issues in disaster with regard to preparedness, recovery and mitigation are addressed by utilizing socio-anthropological and political frameworks. Finally in Part V, a conclusion sums up the findings of the volume and proposes several thoughts on the course of future research.

References

Acker, Joan (1992). "Gendered Institutions: From Sex Roles to Gender Institutions", *Contemporary Sociology*, Vol. 21, pp. 565–569.

Agarwal, B. (1990). "Social Security and the Family: Coping with Seasonality and Calamity in Rural India", *Journal of Peasant Studies*, Vol. 17, No. 3, pp. 341–412.

Agarwal, B. (1992). "The Gender and Environment Debate: Lessons from India", *Feminist Studies*, Vol. 18, No. 1, pp. 119–158.

Alway, Joan, Belgrave, Linda Liska, and Smith, Kenneth J. (1998). "Back to Normal: Gender and Disaster", *Symbolic Interaction*, Vol. 21, No. 2, pp. 175–195.

Amaratunga, Carol A. and O'Sullivan, Tracey L. (2006). "The Path of Disasters: Psychosocial Issues for Preparedness, Response and Recovery", *Prehospital and Disaster Medicine*, May–June, pp. 149–153.

Anderson, Mary B. (1994). "Understanding the Disaster-Development Continuum: Gender Analysis Is the Essential Tool", *Focus on Gender*, Vol. 2, No. 1, pp. 7–10.

Ariyabandu, M. M. (2009). "Sex, Gender and Gender Relations in Disasters", in *Women, Gender and Disaster—Global Issues and Initiatives*, ed. E. Enarson and P. G. D. Chakrabarti. New Delhi: Sage Publications. pp. 5–17.

Begum, R. (1993). "Women in Environmental Disasters: The 1991 Cyclone in Bangladesh", *Focus on Gender*, Vol. 1, No. 1, pp. 34–39.

Blaikie, P., Cannon, T., Davis, I., and Wisner, B. (1994). *At Risk: Natural Hazards, People's Vulnerability, and Disasters*. New York: Routledge.

Bland, L., Brunsdon, C., Hobson, D., and Winship, J. (1978). "Women 'Inside and Outside' the Relations of Production", in *Women Take Issue: Aspects of Women's Subordination*. Women's Studies Group CCCS. London: CCCS.

Bolin, R. C. and Bolton, P. (1986). *Race, Religion and Ethnicity in Disaster Recovery*. Monograph No. 42. Boulder, CO: University of Colorado.

Bolin, R. C., Jackson, M., and Crist, A. (1998). "Gender Inequality, Vulnerability, and Disaster: Issues in Theory and Research", in *The Gendered Terrain of Disaster: Through Women's Eyes*, ed. E. Enarson and B. H. Morrow. Westport, CT: Praeger. pp. 27–43.

Bolin, R. and Stanford, Lois (1998). *The Northridge Earthquake: Vulnerability and Disaster*. London: Routledge.

Bradshaw, S. (2001). *Dangerous Liaisons: Women, Men, and Hurricane Mitch*. Managua, Nicaragua: Fundacion Puntos de Encuentro.

Britton, N. (2001). *A New Emergency Management for the New Millennium?* Key Note Paper presented to the 2nd International Conference on 'Cities on Volcanoes', Auckland, July.

Brody, A., Demetriades, J., and Esplen, E. (2008). *Gender and Climate Change: Mapping the Linkages*. A Scoping Study on Knowledge and Gaps. Brighton: BRIDGE, Institute of Development Studies (IDS).

Cannon, T (1994) "Vulnerability Analysis And The Explanation Of 'Natural' Disasters" in Disasters, Development And Environment, ed. A. Varley. Chichester: John Wiley & Sons, pp. 13–30.

Chant, Sylvia and Gutmann, Matthew C. (2005). "Men-Streaming Gender? Questions for Gender and Development Policy in the Twenty-first Century", in *The Anthropology of Development and Globalization: From Classical Political Economy to Contemporary Neoliberalism,* ed. Marc Edelman and Angelique Haugerud. Oxford: Blackwell. pp. 240–249.

Childers, C. (1999). "Elderly Female-Headed Households in the Disaster Loan Process", *International Journal of Mass Emergencies and Disasters,* Vol. 17, No. 1, pp. 99–110.

Clason, Christine (1983). "The Family as Life-saver in Disaster", *Journal of Mass Emergencies and Disasters*, Vol. 1, No. 1, pp. 43–62.

CSE (Centre for Science and Environment) (1985). *The State of India's Environment 1984–1985: The Second Citizens' Report*. New Delhi: CSE.

Cutter, S (2003) "The Vulnerability of Science And The Science Of Vulnerability", *Annals of the Association of American Geographers,* Vol. 93, No. 1, pp. 1–12.

Dankelman, Irene (2002). "Climate Change: Learning from Gender Analysis and Women's Experiences of Organising for Sustainable Development", *Gender and Development*, Vol. 10, No. 2, pp. 21–29.

Davidson, J. (1993). "Women's Relationship with the Environment", *Focus on Gender*, Vol. 1, No. 1, pp. 5–10.

Denton, F. (2002). "Climate Change Vulnerability, Impacts, and Adaptation: Why Does Gender Matter?", *Gender and Development*, Vol. 10, pp. 10–20.

Doyal, Lesley (2001). "What Makes Women Sick", in *The Sociology and Politics of Health: A Reader*, in Michael Purdy and David Banks. London: Routledge. pp. 88–95.

Drabek, Thomas E. (1969). "Social Processes in Disaster: Family Evacuation", *Social Problems*, Vol. 16, No. 3, pp. 336–349.

Enarson, E. (1998). "Through Women's Eyes: A Gendered Research Agenda for Disaster Social Science", *Disasters*, Vol. 22, No. 2, pp. 157–173.

Enarson, E (1999) "Women and Housing Issues in Two U.S. Disasters", *International Journal of Mass Emergencies and Disasters*, Vol. 17, No. 1, pp. 39–63.

Enarson, E. (2000). *Gender Issues in Natural Disasters: Talking Points and Research Needs*. ILO InFocus Programme on Crisis Response and Reconstruction Workshop, Geneva, May 3–5.

Enarson, E. (2012). *Women Confronting Natural Disaster: From Vulnerability To Resilience*. Boulder/London: Lynne Rienner Publishers.

Enarson, E. and Fordham, M. (2001). "From Women's Needs to Women's Rights in Disasters". *Environmental Hazards*, Vol. 3, pp. 133–136.

Enarson, E. and Fordham, M. (2004). "Lines That Divide, Ties That Bind: Race, Class, and Gender in Women's Flood Recovery in the U.S. and U.K.". *Australian Journal of Emergency Management,* Vol. 15, No. 4, pp. 43–52.

Enarson, E. and Morrow, B. (1997). "A Gendered Perspective: The Voices of Women", in *Hurricane Andrew: Ethnicity, Gender and the Sociology of Disasters*, ed. W. Peacock, B. Morrow, and H. Gladwin. New York: Routledge. pp. 116–140.

Enarson, E. and Morrow B. H. (eds.) (1998). *The Gendered Terrain of Disaster: Through Women's Eyes*. Westport, CT: Greenwood Publications.

Enarson, Elaine, Fothergill, Alice and Peek, Lori (2007). "Gender and Disaster: Foundations and Directions", in *Handbook of Disaster Research*, ed. Havidán Rodríguez, Enrico L. Quarantelli and Russell R. Dynes. New York: Springer. pp. 130–146.

Enarson, E. and Phillips, B. (2008). "Invitation to a New Feminist Disaster Sociology: Integrating Feminist Theory and Methods", in *Women and Disasters: From Theory to Practice*, ed. B. Phillips and B. H. Morrow. Philadelphia, PA: Xlibris, International Research Committee on Disaster. pp. 41–74.

Fernando, P. and Fernando, V. (1997). *South Asian Women Facing Disasters, Securing Life*. Colombo, Sri Lanka: ITDG.

Fothergill, A. (1996). "Gender, Risk, and Disaster", *International Journal of Mass Emergencies and Disasters*, Vol. 14, No. 1, pp. 33–56.

Fothergill, A. (2004). *Heads above Water: Gender, Class, and Family in the Grand Forks Flood.* Albany, NY: State University of New York Press.

Fothergill, A., Maestas, E. G. M., and Darlington, J. D. (1999). "Race, Ethnicity, and Disasters in the United States: A Review of the Literature", *Disasters*, Vol. 23, No. 2, pp. 156–173.

Fritz, Charles E. and Marks, Eli S. (1954). "NORC Studies of Human Behaviour in Disaster", *Journal of Social Issues*, Vol. 10, No. 3, pp. 26–41.

Gardner, Katy and Lewis, Davis (1996). *Anthropology, Development and the Post-modern Challenge*. London: Pluto Press.

Ginige, Kanchana, Amaratunga, Dilanthi, and Haigh, Richard (2009). "Mainstreaming Gender in Disaster Reduction: Why and How?", *Disaster Prevention and Management*, Vol. 18 No. 1, pp. 23–34.

Glazer, Nona (1980). "Everyone Needs Three Hands: Doing Unpaid and Paid Work in Berk", in *Women and Household Labour*, ed. F. Sarah. Beverly Hills, CA: Sage Publications. pp. 249–270.

Hewitt, K. (1983). *Interpretations of Calamity from the Viewpoint of Human Ecology*. London: Allen and Unwin.

Hewitt, K. (1997). *Regions of Risk: A Geographical Introduction to Disasters*. Reading, MA: Addison Wesley Longman.

IFRC (International Federation of the Red Cross and Red Crescent Societies) (2003). *World Disasters Report 2003*. Oxford: Oxford University Press.

Ikeda, K. (1995). "Gender Differences in Human Loss and Vulnerability in Natural D: A Case Study from Bangladesh", *Indian Journal of Gender Studies*, Vol. 2, No. 2, pp. 171–193.

International Strategy for Disaster Reduction (2004). "Terminology: Basic Terms of Disaster Risk Reduction". Available at www.unisdr.org/eng/library/lib-terminology-eng%20home.htm (Accessed June 18, 2007).

Johnsson-Latham, G. (2007). *A Study on Gender Equality as a Prerequisite for Sustainable Development*. Stockholm, Sweden: Environment Advisory Council.

Kafi, S. A. (1992). *Disaster and Destitute Women: Twelve Case Studies*. Dhaka: Disaster Resource Unit, Bangladesh Development Partnership Centre.

Kelkar, K. and Nathan, D. (1991). *Gender and Tribe: Women, Land and Forests in Jharkhand*. New Delhi: Kali for Women.

Khondker, H. H. (1996). "Women and Floods in Bangladesh", *International Journal of Mass Emergencies and Disasters*, Vol. 14, No. 3, pp. 281–292.

Lorber, Judith. (1994). *Paradoxes of Gender*. New Haven, CT: Yale University Press.

Lorber, Judith (1998). *Gender Inequality: Feminist Theories and Politics*. Los Angeles: Roxbury.

McEntire, D. A. (2001). "Triggering Agents, Vulnerabilities and Disaster Reduction: Towards a Holistic Paradigm", *Disaster Prevention and Management*, Vol. 10, No. 3, pp. 189–196.

Morrow, B. H. and Enarson, E. (1996). "Hurricane Andrew through Women's Eyes: Issues and Recommendations", *International Journal of Mass Emergencies and Disasters*, Vol. 14, No. 1, pp. 1–22.

Myers, Mary (1994). "'Women and Children First': Introducing a Gender Strategy into Disaster Preparedness", *Focus on Gender*, Vol. 2, No. 1, pp. 14–16.

Neumayer, Eric and Plumper, Thomas (2007). "The Gendered Nature of Natural Disasters: The Impact of Catastrophic Events on the Gender Gap in Life Expectancy, 1981–2002", *Annals of the Association of American Geographers*, Vol. 97, No. 3, pp. 551–566.

Noel, G. E. (1998). "The Role of Women in Health-related Aspects of Emergency Management: A Caribbean Perspective", in *The Gendered Terrain of Disaster: Through Women's Eyes*, ed. E. Enarson and B. H. Morrow. Westport, CT: Praeger. pp. 213–223.

Oldrup, H. and Breengaard, M. H. (2009). *Desk Study on Gender, Gender Equality, and Climate Change*. Nordic Council of Ministers.

Ollenburger, J. C. and Tobin, G. A. (1998). "Women and Post-disaster Stress", in *The Gendered Terrain of Disaster: Through Women's Eyes*, ed. E. Enarson and B. H. Morrow. Greenwood, CT: Praeger. pp. 95–108.

Oxfam (2005). *The Tsunami's Impact on Women*. Oxfam Briefing Note.

Parida, P. K. (2002). "Towards Rebuilding a Post-Disaster Society: A Case Study of Super-Cyclone Affected Coastal Orissa", *The Indian Journal of Social Work*, Vol. 63, No. 2, pp. 243–262.

Parida. P. K. (2008). "Super Cyclone Affected Coastal Orissa: A Social Vulnerability Approach", *Review of Development and Change*, Vol. XIII, No. 2, pp. 159–180.

Parida, P. K. (2009). "Vulnerability of Disaster Victims: A Study from Coastal Orissa", *Eastern Anthropology*, Vol. 62, No. 2, pp. 209–224.

Parida, P. K. (2010). "Understanding Evacuation Behaviour in a Disaster Society: The Case of Coastal Orissa", *Sociological Bulletin*, Vol. 59, No. 2, pp. 179–198.

Parida, P. K. (2014) "'We are Living with Daily Disasters': A Study of Mental Health Consequences of Cyclone 'Thane'", *Sociological Bulletin*, Vol. 63, No. 1, pp. 77–94.

Parida, P. K. (2015a). "The Social Construction of Gendered Vulnerability to Tsunami Disaster: The Case of Coastal Sri Lanka", *Journal of Social and Economic Development*, Vol. 17, No. 2, pp. 200–222.

Parida, P. K. (2015b). "Natural Disaster and Women's Mental Health", *Social Change*, Vol. 45, No. 2, pp. 256–275.

Parida, P.K.(2016). "Understanding the Elderly in Natural Disaster: Need for Comprehensive Disaster Planning", *Research Process*, Vol. 4, No. 1, pp. 51–62.

Peacock, W. G. and Ragsdale, A. K. (1997). "Social Systems, Ecological Networks, and Disasters: Toward a Socio-Political Ecology of Disasters", in *Hurricane Andrew: Ethnicity, Gender, and the Sociology of Disasters*, ed. W. G. Peacock, B. H. Morrow, and H. Gladwin. New York: Routledge. pp. 20–35.

Peacock, W. G., Morrow, B. H., and Gladwin, H. (1997). *Hurricane Andrew: Ethnicity, Gender and the Sociology of Disaster*. London: Routledge.

Peuler, J. N. (1988). "Community Outreach after Emergencies", in *Mental Health Response to Mass Emergencies: Theory and Practice*, ed. M. Lystad. New York: Brunner/Mazel. pp. 239–261.

Phillips, B. (1990). "Gender as a Variable in Emergency Response", in *The Loma Prieta Earthquake: Studies of Short-Term Impacts*, ed. R. Bolin. Boulder, CO: Natural Hazards Research and Applications Information Center, University of Colorado. pp. 84–90.

Quarantelli, E (1990) "Disaster Prevention and Mitigation in Lada". Paper presented at Colloquium on the Environment and Natural Disaster Management, Washington, DC: World Bank.

Rivers, J. W. (1982). "Women and Children Last: An Essay on Sex Discrimination in Disasters", *Disasters*, Vol. 6, No. 4, pp. 256–267.

Rocheleau, D., Thomas-Slayter, B., and Wangarai, E. (eds.) (1996). *Feminist Political Ecology: Global Issues and Local Experiences*. New York: Routledge.

Rubin, Gayle. (1975.) "The Traffic in Women" in *Toward an Anthropology of Women*, ed. R. Reiter. New York: Monthly Review Press. pp. 157–211.

Sen, A. (1981). *Poverty and Famines: An Essay on Entitlement and Deprivation*. Oxford: Oxford University Press.

Sen, A. (1988). "Family and Food: Sex Bias in Poverty", in *Rural Poverty in South Asia*, ed. T. Srinivasan and P. K. Bardhan. New York: Columbia University Press. pp. 453–472.

Sen, A. (1990). "Gender and Cooperative Conflict", in *Persistent Inequalities: Women and World Development*, ed. I. Tinker. Oxford: Oxford University Press. pp. 123–149.

Sen, A. (2000). *Development as Freedom*. New York: Oxford University Press.

Shiva, V. (1988). *Staying Alive: Women, Ecology and Development*. New Delhi: Zed Books/Kali for Women.

Shiva, V. (1991). *Ecology and the Politics of Survival-Conflicts over Natural Resources in India*. Delhi: Sage Publications.

Steady, F. C. (ed.) (1993). *Women and Children First: Environment, Poverty, and Sustainable Development*. Rochester, VT: Schenkman Books.

Swaminathan, Padmini (1987). "State and Subordination of Women", *Economic and Political Weekly*, Vol. 22, No. 44, pp. WS34–WS39.

Tinker, I. (ed.) (1990). *Persistent Inequalities: Women and World Development*. Oxford: Oxford University Press.

United Nations (1991). *The World's Women 1970–1990: Trends and Statistics*. Social Statistics and Indicators, Series, K, no. 8 (New York: UN).

United Nations (2009). *Making Disaster Risk Reduction Gender Sensitive*. Geneva: UNISDR, UNDP and IUCN.

UNISDR (United Nations Office for Disaster Risk Reduction) (2004). "*Living with Risk: A Global Review of Disaster Reduction Initiatives*", Available at www.unisdr.org

UNRISD (The United Nations Research Institute for Social Development) (1992). *Social Development Research, 1991–92*. Geneva: UNRISD.

WHO (World Health Organization) (1992). *Women's Health: Across Age and Frontier*. Geneva: WHO.

WHO (2002). *Gender and Health in Disasters*. Geneva: Department of Gender and Women's Health.

WHO (2015). *Gender* [Online]. Available at www.who.int/topics/gender/en/ (Accessed May 21, 2015).

Wiest, R. E. (1998). "A Comparative Perspective on Household, Gender, and Kinship in Relation To Disaster", in *The Gendered Terrain of Disaster: Through Women's Eyes*, ed. E. Enarson and B. H. Morrow. Westport, CT: Praeger. pp. 63–80.

Wisner, B., Blaikie, P., Cannon, T., and Davis, I. (2004). *At Risk: Natural Hazards, People's Vulnerability, and Disaster* (2nd ed.). London: Routledge.

World Bank (1993). *World Development Report 1993: Investing in Health*. Oxford: Oxford University Press.

Yelvington, K. (1997). "Coping in a Temporary Way: The Tent Cities", in *Hurricane Andrew: Ethnicity, Gender, and the Sociology of Disaster*, ed. W. Peacock, B. Morrow, and H. Gladwin. New York: Routledge. pp. 92–115.

PART II

Gender Inequality, Susceptibility and Disaster

2

Gender and Disaster

An Overview of Differential Vulnerabilities

Sudha Sitharaman

A very severe cyclonic storm, 'Thane', engulfed the Union Territory of Puducherry in southeast India on 29 and 30 December 2011. The fatal event caused damage to various sectors besides killing 16 people in Puducherry alone. Due to the heavy rain accompanied by whirlwind, seawater entered several settlements in the coastal region forcing the residents to move to safer places. Sustained winds of 140 miles per hour with gusts to 180 were recorded. On the morning of 31 December, as I sat in my house in Pondicherry University campus, Cyclone Thane had left at least 46 dead in Tamil Nadu and Puducherry. Cuddalore and Puducherry were worst affected by the cyclone. Houses were left without most if not all of their roofs. The streets were full of debris, roofs, clothes, street lights, telephone and power poles. The storm made Cuddalore inaccessible by damaging road network, and the National Disaster Respond Force (NDRF) and fire and rescue services had difficulty in reaching the cyclone-affected fishing hamlets. The cyclone severely damaged the power network in Cuddalore district of Tamil Nadu and Puducherry, and it is estimated that 4,500 transformers were damaged in Cuddalore district alone. Fishing activities had come to a complete halt in the cities of Chennai, Thiruvallur and Kanchipuram in the wake of the storm after the warnings by the weather department were received. Approximately

10,000 fishing boats were moved to safer locations in all fishing hamlets in these areas.

Disasters have been on the rise and are predicted to increase in frequency and intensity due to climate change, rapid urbanization and environmental degradation. Exposure to disasters may be direct, involving personal loss, or indirect, because of socio-economic conditions of the vulnerable community, shared community damage and disruption (Morrow 1999). Studies on the impact of disasters suggest that as extraordinary events, they disrupt social organization and radically interfere with the normal flow of events (Olesen 1992). Impact of disasters on health indicate that they result in severe, lasting and comprehensive psychological effects which are also characterized as causing a high prevalence of injury, threat to or loss of life, extreme and widespread damage to property, and serious financial problems for the affected community (Norris et al. 2002). Nevertheless, no matter how temporary the destruction and disruption, disasters impose the necessity of re-orienting and reorganizing responses to the primary world of family and neighbours as well as the secondary world of community relationships and organization (Goode 1992; Kreps 1984; Turner et al. 1986).

How does one understand the experiences of victims when the world as one knew it is destroyed? The 'non-everyday' qualities of disasters make them important research settings. This, however, is not to argue that we keep the processes of disaster away from the social structures of the 'everyday': To draw from the everyday life approaches of interpretive sociology (Adler et al. 1987) and use the concepts of the 'everyday world' and 'taken-for-granted reality' (Berger and Luckmann 1966) in our understanding of these experiences. Everyday life approaches in sociology—whether from the phenomenological or the symbolic interactionist traditions—examine the ongoing, active processes through which people define and impute meaning to self, others and social situations. These approaches ask how people interpret, understand and construct the interactions in which they are routinely involved and how such interpretations then influence their self-concepts and subsequent actions. Individuals are treated as conscious, reflexive and creative participants in their social worlds and the socially constructed 'stock of knowledge' (Schutz 1970) that defines physical and social reality. Employing these concepts helps us to identify phases in the disruption and reconstruction of the everyday world and to identify the techniques people used for operating in an everyday world that no longer worked. Indeed, disasters are inherently social phenomenon embedded in the social structure linked to who we are, how we live our lives and how we structure and maintain our society. While the newer approaches to disasters locate its roots in social structure, there is a large ground not covered by sociological approaches. Research in the field has amply made clear that disasters—whether predictable or unpredictable, natural or socially constructed—do not affect people indiscriminately.

In research concerning gender, inclusion has largely remained at the level of mapping basic information on gender differences, as demographic variable. Typically, women, remain as unitary category, are contrasted to men, likewise unitary category, refusing to take on board the intersectionalities of caste, class, religion and region. Gender dimension of social structure is happy inclusion into disaster research, critical engagement through explanation and analysis on women's experiences and perspectives in disaster requires further exploration. Amidst debates that problematize the category of 'woman' as the subject of feminist politics (for e.g. Menon 2004) and emphasize the intersectionality of gender with class, caste, religion and ethnicity, it would be fruitful to examine the intersectional nature of multiple forms of inequalities in the context of disasters. Thus, the need is for an inclusive approach to vulnerability reduction, and resilience building has never been more urgent. What follows in the chapter is an overview of the socio-political events of disaster as social phenomena.

Overview of Gender in Disasters

Gender issues are not manifestations of disaster-related crisis situations, but are prevalent in society, operational and visible in daily life at the level of the individual, family, community and society at large and reflected institutionally as well as in social and cultural norms (Ariyabandu 2009: 5). The sociological study of gender has moved through a number of phases in the last three decades. Traditionally, social differences between women and men have been viewed as the natural consequence of biological differences. Maleness and femaleness were looked upon as biologically given properties of individuals with clear implications for social behaviour. However, explanations of sex and sexuality that assume that the meaning of women's social existence can be derived from facts of their physiology were questioned. The distinction between sex and gender helped feminist scholars to move discussion away from this 'sex differences' approach by presenting gender as a social construction rather than a biological construct. Sociological attention then began to focus on gender (or sex) roles and socialization. In the more static, functionalist approach to gender roles, the emphasis was on the acquisition and enactment of normatively defined patterns of masculine and feminine behaviour. Gender is omnirelevant (West and Zimmerman 1987). Male and female are not simply roles we can assume or relinquish; we must always enact (however imperfectly or creatively) one or the other. Furthermore, a focus on gender roles obscures the institutional and structural (including hierarchical) aspects of gender. Roles and socialization do not tell us much, for example, about the gendered organization of the economy and the family or why diverse and often inconsistent behavioural expectations, outcomes and social values are linked to the categories of female and male. Gender is now increasingly understood as a property not of individuals or interaction but of social organization; it is an aspect of social structure, not an individual attribute or role (see, for example, Acker 1992; Lorber 1994; Rubin 1975). Like caste and class, gender is a social relationship of difference and inequality that is omnirelevant precisely because it organizes and affects all dimensions of social life.

Gender not only organizes institutions and identities, but also shapes language and knowledge and even the ways bodies are known, experienced, used and regulated. Understanding gender as a structure that cuts across and organizes all dimensions of social life provides a means for investigating and understanding the links between different dimensions or 'levels' of social reality. As gender structures knowledge and language, social institutions, interaction, identities and bodies, it also connects them and provides a means of translation and transmission between levels. Because it organizes all dimensions of social life, gender also provides a means for translating ideas into actions, culture into institutional practices and bodily experiences into identity. At the same time, gender serves as a means of transmission. Here, the conceptualization of gender as a social relationship of difference and inequality that organizes and affects all dimensions of social life, rather than a role, is especially significant in that it allows struggle (between individuals or groups) to be as significant a mode of transmission as socialization or negotiation. This understanding of gender allows us to grasp not only the fluidity and contested nature of gender roles but also their resilience and durability. Among different groups, in different interactional contexts, and under different circumstances, the expectations and behaviours associated with being male and female may vary and change, but we remain recognizably to ourselves and other women and men, girls and boys. Even when everyday life is disrupted, when most familiar activities and settings are upset or destroyed, organizing principles such as gender remain operative not only in the roles and identities that shape our behaviour and interaction but also in the institutional arrangements that bring back normal routines, in how one learns about and makes sense of what has happened, and in the ways one experiences and uses one's body. In distinguishing sex from gender, feminist theorists have disputed causal explanations that assume that sex dictates or necessitates certain social meanings for women's experience. Feminist theory has sought to understand the way in which systemic or pervasive political and cultural structures are enacted and reproduced through individual acts and practices, and how the analysis of ostensibly personal situations is clarified through situating the issues in a broader and shared cultural context. If gender is not recognized as socially constructed, historically variable and a primary dimension of frequently vast symmetries in social and economic power, gender-based inequalities inevitably will be ignored (Ortner and Whitehead 1981).

Gender-based prejudices and divisions in many societies mainly affect girls and women and as these are based on views of them as physically and emotionally weak, inferior in comparison to men and boys, dependent, subordinate and a burden to family. Such gender-based perceptions form the nature of interactions at personal, family and institutional levels and contribute towards the formulation of gendered attitudes leading to observations, decisions and actions within family and society, as well as in the formal institutions of different kinds including the state. The ideal of the heterosexual, gender-role differentiated family, in particular, has come under severe criticism by feminists. Liberal feminists such as Betty Friedan argued that the husband as breadwinner/wife as homemaker model of family has led to women's subjugation in the household. Radical feminists argue that institutions such as marriage and motherhood are oppressive as they essentialize women and are instrumental in their continual subjugation (Ferguson 2007; Beauvoir 1953). Within the Indian context, Nivedita Menon (2012) argues that the patriarchal heterosexual family as an institution, is based on inequality and functions as the core of the social order. The normalization of the heterosexual patriarchal family, she argues, is also inextricably linked to other axes of oppression such as caste, race and community. For example, it is recognized today that caste purity and patrilineal succession are ensured through the strict regulation of women's sexuality through endogamous marriages, without which the caste order cannot be sustained. Gender relations come into effect in all spheres of life—personal, social, economic and political; they are not equitable and almost always unfavourable and prejudiced towards girls and women. Besides cultural patterns structure the lives of women and men to differing needs, roles and social power in different social and cultural contexts. Men are usually seen as primary income generators, while women's economic activities, often the mainstay of the household economy, are less visible. Women carry the primary responsibility for the care of children, the elderly, the disabled and the ill. Women and men of varying age groups have different life experiences and skills owing to gendered identities, roles and responsibilities. According to Ariyabandu (2009), gender relations in society are broadly reflected in *gendered identities*—a combination of physical and behavioural characteristics which set apart boys from girls, men from women; *perceptions*—views as to how they are differentiated in their roles as men and women; *attitudes*—actions guided by the perceptions; and *status*—the place occupied by men and women in family, community and society (p. 6). Gender-based social, economic, religious and cultural constructs marginalize women across all community groups irrespective of class, caste, economic standing, status, ethnicity, age and so on. Gender relations are thus reproduced in a manner such that women come to occupy a subordinate status within family, community and society in comparison to their male counterparts. Bourdieu (1977) has noted that gender relations are organized everywhere by dominant power relations and play themselves out in everyday life in such a way that they appear natural and inevitable to the participants. This is key to understanding gender-based differences in disaster impact, response and recovery.

Differential Vulnerabilities

Perhaps the most important lesson learned from disaster sociology is that social vulnerability to disaster is not uniform or universal. Historically, the dominant theoretical perspectives, research strategies and guiding questions in disaster social science have been determinedly male-oriented if not male-dominated (Bolin et al. 1998; Enarson and Morrow 1998; Enarson and Phillips 2008; Hewitt 1998). Everyday life approaches offer provocative insights into socially constructed meanings and actors' choices, but they strain to connect interaction to larger social contexts, dynamics and hierarchies. As Nicos Mouzelis (1993) has observed, everyday life forms of sociology 'tend to forget that actors, because of their very unequal access to the economic, political and cultural means of production, contribute unequally to the construction of social reality' (p. 679). As ever, context is the key. Besides, the social vulnerability approach does not focus on adaptation to hazard but on the complex development processes undermining disaster resilience and putting rising numbers of people at increased risk (among others, see Blaikie et al. 2004; Bolin and Stanford 1998; Hewitt 1997, 1998; Holloway 1999; Mileti 1999).

Disasters affect women and men differently, and due to deep-seated gender inequalities, women are at greater risk of suffering from disasters. As one is aware, world over, deprived access to education, health and food resources has hindered mobility and decision-making capacities, subjecting them to legal, political, institutional and market inequalities and lower level of human rights, broadly defined, than men (Bolin et al. 1998; Enarson 2004; Fothergill 1998). Compounding the predetermined differentials is burdensome gender roles, especially the role as 'caretaker' and 'nurturer of the family', and the overall lesser value attributed to women in society, which jointly serve to enhance women's jeopardy in disasters (Enarson 2000; Peek and Fothergill 2008). Collectively, these socially determined marginalities combine to form a complex web of predisposed female vulnerability to disasters, as stated by Enarson and Morrow:

> Gender is a pervasive division affecting all societies, and it channels access to social and economic resources away from women and towards men. Women are often denied the right to vote, the right to inherit land and generally have less control over income-earning

> opportunities and cash within their own households ... in the absence of other compensations to provide safe conditions, (this) leads to increased vulnerability, (therefore), we contend that in general women are more vulnerable to disasters. (Enarson and Morrow 1998: 2)

Pre-existing vulnerabilities reveal themselves and submit persons to heightened impact depending on geography (e.g., if the setting is near a river, coast or fault line), age (e.g., children and the elderly) and socioeconomic status (Juran 2012: 3). Discrimination against women does not only accentuate women's vulnerabilities during disasters, but also waste women's potential as sources of resilience.

Scholars of disaster gender vulnerabilities have empirically documented the disproportional effects of disasters upon women pan-globally (Ahsan and Khatun 2004; Enarson 2000; Enarson and Morrow 1998). Paradoxically, vivid accounts from the field in particular disaster events frequently document disproportionate impacts on women and girls with respect to health and security, livelihood, social power, human rights and, too often, life itself. Women's roles in securing water, food and shelter, during and after disasters, have been well documented. Yet women are still marginalized in decision making on disaster issues, even as they often hold vital social knowledge and vast untapped capacity for reducing community risk. In her review of the literature on gender and disaster, Fothergill (1996) notes the importance of women's household management skills in disaster situations and the primary social world of family and neighbours. Fothergill and others (2014) (also see Anderson and Manuel 1994; Johnston and Johnson 1989) find significant gender differences in perceptions of risk; activities before, during and after the disaster; and reactions to stress. Many researchers find that females express more mental health problems from disasters, such as stress, depression and posttraumatic stress disorder. As Parida (2014: 84) notes in his study of cyclone Thane in Puducherry, women experience greater stress than men in disaster situations. The psychological stressors of the event as well as pre-existing differential social stressors such as poverty, hunger, malnutrition, bad living conditions and inadequate health care services contributes to post-disaster mental health consequences. The heightened exposure results from their social caste, class and their caregiving roles, and their relative lack of power and status. In disaster situations, women have to stay with, assist, protect and nurture family members.

In this new paradigm, gender is not seen as incidental or derivative but contextualized in relation to other social markers of dominance and subordination, understanding women and men (but women far more than men) as embodied and embedded gendered actors in a gender ordered world. The new approach to gender that we see on the horizon also emphasizes the contradictions in women's lives and their different and sometimes divergent needs and interests in disaster contexts. Perhaps it is this phenomenon that has led Enarson and Morrow (1998) to claim that there exist 'gendered disaster vulnerabilities' and, therefore, a 'gendered terrain of disasters' (p. 8). Thus, while the post-disaster situation varies greatly from the 'normal' time/space of the 'everyday', the overarching parallel is that disparities that existed before the disaster are perpetuated and exacerbated both during and after the disaster. The post-disaster period thus forms a continuum in which pre-existing gender inequalities are maintained and reproduced, if not magnified. In fact, the conclusion to be drawn from the disasters of the past two decades is that sex and gender are never automatically the primary social facts on the ground nor are these ever in play in isolation from other facts of life. Although substantial barriers exist in a gender equitable approach when we confront entrenched male dominance as a root cause of social vulnerability, at the turn of the twenty-first century, we also note the counter trends: A new community of practice is emerging based on converging interests, inter-organizational coalitions and interpersonal networks leading to gender and disaster specialists, activists, advocates and practitioners no longer working in isolation. Women's survival and coping skills, their interpersonal networks and intimate care of the most vulnerable and certainly their knowledge of environmental resources and ecosystems are all lifesaving.

References

Acker, Joan (1992). "Gendered Institutions: From Sex Roles to Gender Institutions", *Contemporary Sociology*, Vol. 21, pp. 565–569.

Adler, Patricia A., Adler, Peter, and Fontana, Andrea (1987). "Everyday Life Sociology", *Annual Review of Sociology*, Vol. 13, pp. 217–235.

Ahsan, R. and Khatun, H. (2004). "Disaster and the Silent Gender", in *Disaster and the Silent Gender: Contemporary Studies in Geography*, ed. R. Ahsan and H. Khatun. Dhaka: The Bangladesh Geographical Society. pp. 1–8.

Anderson, Karen N. and Manuel, Gerdenio (1994). "Gender Differences in Reported Stress Response to the Loma Prieta Earthquake", *Sex Roles*, Vol. 30, pp. 725–733.

Ariyabandu, M. Malalgoda. (2009). "Sex, Gender and Gender Relations in Disasters", in *Women, Gender and Disaster: Global Issues and Initiatives*, ed. Elaine Enarson and P. G. Dhar Chakrabarti. Delhi: Sage Publications. pp. 5–17.

Beauvoir, Simone de. (1953). *The Second Sex*. London: Knopf.

Berger, Peter and Luckmann, Thomas (1966). *The Social Construction of Reality*. New York: Doubleday.

Bolin, R. and Stanford, L. (1998). *The Northbridge Earthquake: Vulnerability and Disaster*. London: Routledge.

Blaikie, T. Cannon. Wisner, B.P. and I. Davis. (2004). *At Risk: Natural Hazards, People's Vulnerability and Disasters* (2nd edition). London and New York: Routledge.

Bolin, R., Jackson, M. and Crist, A. (1998). Gender Inequality, Vulnerability, and Disaster: Issues in Theory and Research. In E. Enarson and B. Morrow (Eds), *The Gendered Terrain of Disaster: Through Women's Eyes* (pp. 27–43). Westport: Praeger Publishers.

Bourdieu, Pierre (1977). *Outline of a Theory of Practice*. Britain: Cambridge University Press.

Enarson, E. (2000). *Crisis Response and Reconstruction: Gender and Natural D*. Working Paper 1. Geneva: Recovery and Reconstruction Department, InFocus Programme.

Enarson, E. (2004). "International Perspectives on Gender and Disaster: Differences and Possibilities", *International Journal of Sociology and Social Policy*, Vol. 24, Nos. 10/11, pp. 49–93.

Enarson, E. and Morrow, B. (1998). "Why Gender? Why Women'? An Introduction to Women and Disaster", in *The Gendered Terrain of Disaster: Through Women's Eyes*, ed. E. Enarson & B. Morrow. Westport: Praeger. pp. 1–8.

Enarson, Elaine and Morrow, Betty Hearn (1998). *The Gendered Terrain of Disaster: Through Women's Eyes*. Westport, CT: Greenwood Publications.

Enarson, Elaine and Phillips, Brenda (2008). "Invitation to a New Feminist Disaster Sociology: Integrating Feminist Theory and Methods", in *Women and Disasters: From Theory to Practice*, ed. Brenda Phillips and Betty Hearn Morrow. Philadelphia, PA: International Research Committee on Disaster, Xlibris. pp. 41–74.

Ferguson, A. (2007). Gay Marriage: An American and Feminist Dilemma. *Hypatia*, pp. 39–57.

Fothergill, A. (1996). 'Gender, Risk and Disaster', *International Journal of Mass Emergencies and Disasters*, 14(1): 33–56.

Fothergill, A. (1998). "Neglect of Gender in Disaster Work: An Overview of the Literature", in *The Gendered Terrain of Disaster: Through Women's Eyes*, ed. E. Enarson and B. Morrow. Westport: Praeger. pp. 11–25.

Fothergill, A. and Peek, Lori. (2014). "Post Disaster Decline: Understanding Children's Vulnerability Before, during, and after Katrina", *The Dialogue, Journal on Disaster Behavioral Health.* Vol 10 (3). pp. 2–4.

Goode, Erich (1992). *Collective Behavior.* Fort Worth, TX: Harcourt.

Hewitt, K. (ed.) (1983). *Interpretations of Calamity: From the Viewpoint of Human Ecology*. London: Allen and Unwin.

Hewitt, Kenneth (1997). *Regions of Risk: A Geographical Introduction to Disasters*. Essex: Longman.

Hewitt, Kenneth (1998). "Excluded Perspectives in the Social Construction of Disaster", in *What Is a Disaster? Perspectives on the Question*, ed. E. L. Quarantelli. New York: Routledge.

Holloway, Ailsa (1999). *Risk, Sustainable Development and Disasters.* Cape Town: Peri Peri Publications.

Johnston, Drue M. and Johnson, Norris R. (1989). "Role Extension in Disaster: Employee Behavior at the Beverly Hills Supper Club Fire", *Sociological Focus*, Vol. 22, pp. 39–51.

Juran, Luke (2012). "The Gendered Nature of Disasters: Women Survivors in Post-Tsunami, Tamil Nadu", *Indian Journal of Gender Studies*, Vol. 19, No. 1, pp. 1–29.

Kreps, Gary A. (1984). "Sociological Inquiry and Disaster Research", in *Annual Review of Sociology*, ed. Ralph H. Turner and James F. Short. Palo Alto, CA: Annual Reviews. pp. 309–330.

Lorber, J. (1994). *Paradoxes of Gender*. New Haven, CT: Yale University Press.

Menon, Nivedita. (2004). *Recovering Subversion: Feminist Politics Beyond the Law.* Delhi: Permanent Black.

Menon, Nivedita. (2012). *Seeing Like a Feminist.* New Delhi: Zubaan.

Mileti, Dennis (1999). *Disasters by Design: A Reassessment of Natural Hazards in the United States*. Washington DC: Joseph Henry Press.

Morrow, B. H. (1999). 'Identifying and Mapping Community Vulnerability', *Disasters*, Vol. 23, No. 1, pp. 1–18.

Mouzelis, Nicos P. (1993). *Back to Sociological Theory: The Construction of Social Orders.* Palgrave Macmillan.

Norris, F. H. Friedman, M. J., and Watson, P. J. (2002). "60,000 Disaster Victims Speak: Part I—An Empirical Review of the Empirical Literature 1981–2001", *Psychiatry*, Vol. 65, No. 3, pp. 7–39.

Olesen, Virginia L. (1992). "Extraordinary Events and Mundane Ailments: The Contextual Dialectics of the Embodied Self", in *Investigating*

Subjectivity, ed. Carolyn Ellis and Michael G. Flaherty. Newbury Park: Sage Publications. pp. 205–220.

Ortner, Sherry B. and Whitehead, Harriet (1981). *Sexual Meanings: The Cultural Construction of Gender and Sexuality*. New York: Cambridge University Press.

Parida, P. Kumar. (2014). "'We Are Living with Daily Disasters': A Study of Mental Health Consequences of Cyclone 'Thane'", *Sociological Bulletin*, Vol. 63, No. 1, pp. 77–94.

Peek, L. and Fothergill, A. (2008). "Displacement, Gender, and the Challenges of Parenting after Hurricane Katrina", *National Women's Studies Association Journal*, Vol. 20, No. 3, pp. 69–105.

Rodríguez, Havidán (2008). "There Is No Such Thing as a Natural Disaster: Race, Class, and Hurricane Katrina", *Social Forces*, Vol. 86, No. 3, pp. 1360–1362.

Rubin, Gayle (1975). "The Traffic in Women" in *Toward an Anthropology of Women*, ed. Rayna Reiter. New York: Monthly Review Press. pp. 157–211.

Schutz, Alfred. (1970). *Reflections on the Problem of Relevance*. New Haven, CT: Yale University Press.

Turner, Ralph H., Nigg, Joanne M. and Paz, Denise Heller (1986). *Waiting for Disaster: Earthquake Watch in California*. Berkeley, CA: University of California Press.

West, Candace and Zimmerman, Don H. (1987). "Doing Gender", *Gender and Society*, Vol. 1, pp. 121–151.

Wisner, B., Blaikie, P., Cannon, T., and Davis, I. (2004). *At Risk: Natural Hazards, People's Vulnerability and Disasters* (2nd edition). London/New York: Routledge.

3

The Socio-cultural Aspects of Gendered Vulnerabilities to Disasters

Case Studies of Women-Headed Households, Post Mumbai Train Blasts, 2006

Nalini Yadav

Introduction

For a long time, studies in the domain of disaster (management) were confined to studying nature of catastrophes, mitigation, preparedness and recovery process (Drabeck and McEntire 2003). Exploring conflict situations and wars and capturing riots were the next development in this field. Social sciences was absent from disaster management research until a new strand of researchers grew up in the early 1990s who conducted studies on culture, religion, gender, race and ethnicity. This paradigm shift emphasized that disasters are strategic sites to study social relationships as disasters strip away the veil that usually compresses social conditions in a very dramatic manner (Fothergill 2004).

Disasters disproportionately affect all members of the society (Neumayer and Plümper 2008). Attributing to their gendered identity, both men and women experience disasters differently. As Mehta (2007) explained that both have distinct roles, responsibilities and differential access to a range of social, economic and political resources. The inequality in accessibility to resources is a key factor behind increased disaster vulnerabilities of women. Enarson (1998) has rightly argued that women experience disasters through intersecting social relations of gender, race and class. The socially determined roles and responsibilities of women and low participation in decision making further increase their pre-existing vulnerabilities.

The chapter examines how unequal gender relations establish gendered vulnerabilities and how these vulnerabilities are reproduced in the aftermath of a disaster, whether natural or man-made. While mentioning categories of disasters, it is important to highlight that in the wake of twenty-first century, terrorist attacks have emerged as another category of disasters. In the past few years, the world has witnessed biological, chemical, radiological, nuclear attacks and explosions (Wough and Sylves 2002). The lethality and destructiveness has disrupted the social order and resulted in intense psychological disorder and emergence of vulnerable households. In view of this, the chapter aims to examine the vulnerabilities of female-headed households after the 2006 train blasts in Mumbai. The concern has increased post-disaster vulnerabilities of female-headed households. This chapter argues that these vulnerabilities are reflections of the unequal gender relations practiced in the social structure.

Literature Review

Studies that have looked into disasters from gender lens have widely observed that before and after a disaster, women become more vulnerable than men of their own class, race, age group and ethnicity. When disasters take away the breadwinner of the family, women are forced to go beyond their predefined role of caregiver which often results in the rise of poor female-headed households (Enarson 1998). The existing literature suggests that post-disaster outmigration and male mortality are leading causes behind the rise of female-headed households (Ariyabandu and Wickramasinghe 2005; Mehta 2007; World Health Organization 2002). According to Moghadam (2005), women are disproportionately living in poverty all over the world and female-headed, single parent households have a poverty rate four times than that of male-headed households. This phenomenon has been referred as *feminization of poverty* (Pearce 1978). Poor individuals, predominantly women and children, are more exposed to hazards and risks. It would be appropriate to say that social structure, gender discriminatory norms and practices further contribute to their plight. For instance, Morrow and Enarson (1996) in their study found that poor minority women in Florida were more at risk of hurricane Andrew because of lack of power, status and resources. Fothergill (1996) has addressed gender issues at various stages of post-disaster recovery. Her study on poverty and disasters in United States suggests that lower income countries have larger number of poor female population with higher mortality rate than men. Thus, poverty is one of the major obstacles that women face in recovery process. It creates lack of employment opportunities and often due to lack of skills, and repressive gender norms, they have to settle for underpaid and unrecognized jobs. Fothergill and Peek (2004) in their review of the *disaster and social class* literature concluded that when a disaster strikes, it is clearly visible that those already living in poverty are impacted in different and significant ways as compared to other members of society. Fordham (1999) further included that poor women naturally take longer time to recover from disasters because of the pre-existing vulnerabilities.

Disasters exacerbate the pre-existing gender discriminatory practices. These discriminations are visible in intra-household distribution of food and relief resources, which instigate women's overall contrasting experience of disasters. Further literature emphasized on repressive gender norms which significantly affect the capacities and mobility of women. Pincha's (2008) extensive study of the Indian Ocean tsunami reflected eye-opening gendered factors. For example, due to rigid gender norms women are not allowed to learn swimming or tree climbing. Since their physical activities and skills are underdeveloped, their bodies are unable to respond in such emergency situations. In addition, women are expected to keep their bodies fully covered with clothes. The consequences of such cumbersome clothing are evident during natural disasters as heavy clothes do not allow them to move swiftly, especially when they need to run away from heavy waves. Adding further to this, Ikeda's (1995) case study on Bangladesh cyclone of 1991 is a rich literature on gendered vulnerabilities during disasters. The study highlighted how the institution of *purdah* (veil) denies women's participation in socio-economic activities and directly prescribes vulnerabilities to disasters. Women often fear the risk of being stigmatized for breaking these sociocultural norms. It results in higher mortality rate, injuries and vulnerabilities of women as compared to men. For example,

Krishnaraj (1997) in her study on Latur earthquake found that in rural areas men sleep in open areas such as outside of the house or on rooftop, which naturally makes it easy for them to quickly evacuate the house as compared to women who sleep in the interiors of the house. Natural disasters present a generalized trend of unequal power which starkly reflects strong patriarchal practices in accessing resources when their distribution is handed over to the existing structure. Further, Ariyabandu and Maithree (2005) added that gender discriminatory practices keep women away from accessing resources in relief and recovery process. Mehta (2007) reflects this in her account of relief and recovery work carried out after devastating floods in northern Pakistan in 2005. Widows and female-headed households found it challenging to access the relief camps that were set up outside the local area. The major concern was to deal with unknown men.

Gender norms and unequal gender relations not only strategically place women as vulnerable population but also do not allow them to solely own property. In disaster recovery process, where reconstructed houses are distributed to survivors, female-headed households find it difficult to claim their ownership if the house was owned by husband or other male member who has passed away in the disaster (Mehta 2007). Enarson (1998, 1999) has agreed in her studies that the pre-existing gender inequality is the core of women's increased vulnerability during disasters. It typically begins at the household level where women have lesser control on resources and employment opportunities which keeps narrowing their opportunities.

The given literature review suggests that while understanding gendered vulnerabilities of disasters, it is significant to look into sociocultural dimension and their impact on widows and female-headed households. The studies conducted so far are limited to the immediate field experiences of humanitarian aid professionals' perspective of gender relations and social changes after natural disasters. Enarson (1998) has also identified this gap and has emphasized on portraits of women-headed households in diverse conditions.

In context of this study, it is important to mention that at present India has been extremely prone to multiple disasters including terrorist attacks. The challenges faced by women in accessing relief and recovery, claiming property and everyday struggle with societal norms, post-natural disasters has been studied however; there is a dearth of literature on similar experiences of single women and female-headed households in the aftermath of terrorist attacks. This study is an attempt to fulfil the existing literature gap and understand the phenomenon of gendered vulnerabilities across man-made, natural disasters and terrorist attacks.

Methods

Context

On the evening of 11 July 2006, a series of seven bomb blasts rocked the city of Mumbai and alarmed the nation with fear. The target was common people, traveling in western line local trains, mostly returning home from work. The trains exploded at or in the vicinity of the western suburban railway stations of Matunga Road, Mahim, Bandra, Khar Road, Jogeshwari, Bhayandar and Borivali.[1] According to Ministry of Railways (2011) report, the railways had no policy with respect to providing jobs to the kin of the deceased in train accidents before this disaster. After the 2006 bomb blasts, it was the first time that Ministry of Railways announced that the kin of those who have died will be considered for Group 'D' posts. Vacancies in this category arise mainly due to retirements, death, promotions, creation of posts and so on, and majority of these jobs involve physical endurance such as janitor and toilet cleaner.

Methodology

The theoretical essence of the study lies in *liberal feminism* which has emerged as a theoretical foundation for contemporary disaster research aimed at examining gendered vulnerabilities (Quarantelli et al. 2007). Liberal feminism is concerned about denial of equal rights to women in education and employment (Walby 1990). It follows that the challenge of the liberal state is to create conditions guaranteeing all women equality of opportunity in social institutions and political life (Enarson and Phillips 2008).

As a case study, the study embraced qualitative research design as qualitative research opens up more to women's voices and needs in general and allows those voices to be heard (Flick 2009). This chapter draws phenomenology as strategy of inquiry. Gillian Rose (1993) in her study, 'Feminism and Geography', established phenomenology as a method to fulfil gaps in research of those disciplines which were earlier not considered important because they had little to do with the public and patriarchal world of geography. Similarly, the chapter delves into the sociocultural milieu of women, further exploring the impacts of a terrorist attack. Based on purposive sampling, a sample size of nine respondents was selected. To bring out maximum information from the respondents, in-depth interviews were conducted. The interviews were further bifurcated in closed-ended and unstructured interview. Closed-ended interviews were helpful in collecting basic details of the respondents along with details on the family. The details included age, education, occupation and income of the family members. The other set of interviews were unstructured, focusing on dialogues and conversation with respondents. The respondents were free to share their experiences of the particular incident and how it unfolded unexpected challenges in their lives.

Sampling

The data collection process for this study went for four months. A right to information (RTI)[2] was filed to the Western Railways, Mumbai, asking for the list of people who were offered jobs by the railways, after the 2006 train blasts. After getting the list, female employees who took the compensation jobs were contacted. In total twenty-five women were contacted but only nine of them agreed to participate in the study. Rests of them were not willing to relive their memories of that period of life, hence they denied participating in the study.

In the first phase, the respondents were introduced with the purpose of study and their consent for participating in the research was taken. Basic information on women and their families, details of the bomb blast and their current employment status were collected. The second phase was to go back to respondents and observe them and the surroundings deeply. In these interviews, participants were more open regarding sensitive issues such as relationship with the deceased, changes in family ties after the disaster, treatment from the family and society, cultural and religious practices and economic status of their family. The last phase was to fulfil the existing gaps in the data and then go back to the respondents for collecting more information. Some of the participants personally gave the interviews, while some of them preferred telephonic interviews due to their prior engagements.

Data Analysis

The study opted for content analysis methods however; the study aims at rich qualitative interview, and therefore, initial analysis was conducted into the following three steps:

Open coding (Labelling the data and creating categories): Strauss and Corbin (1998) have illustrated three methods for conducting data analysis. Open coding is the first step of data analysis in the study. The data was closely observed over and again and line by line analysis was conducted. There were certain facts and contexts in the interviews which ultimately lead to the objectives of the study. Each phenomenon which was a significant indication for the study was labelled with a specific code. This process is known as open coding of data.

Axial coding (Rationally linking the categories with each other): The second step of data analysis is axial coding which explores relations between the categories. The process basically emphasizes on causal relationship between categories. In this process, Strauss and Corbin (1998) suggested the following model to establish relationship between the categories (Figure 3.1).

Figure 3.1

A basic model of axial coding by Strauss and Corbin (1998)

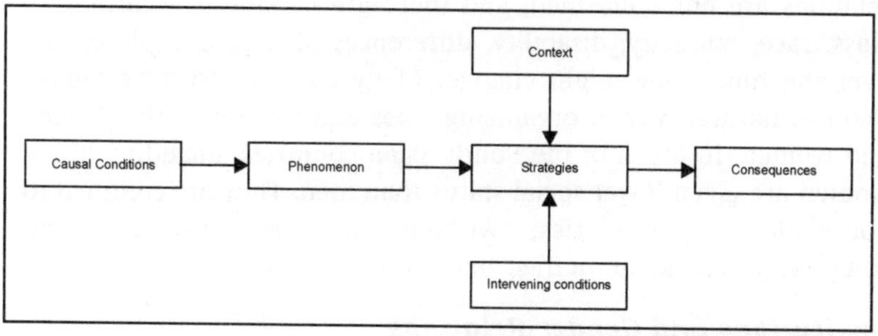

Causal conditions are the active variables such as events that lead to the development of phenomenon. Phenomenon refers to the concept or subject of the study. Context refers to the background variables or moderating variables that hold specific locations in the study. Intervening conditions are mediating variables, although researchers have not established distinctions between context and intervening conditions. Strategies refer to the action taken by agents in response to phenomenon and intervening conditions. Finally, the model leads to the unintended and intended consequences of the actions.

Selective coding: The third and final stage of data analysis was selective coding. In this process, amongst all the categories, a core category is selected which has the capacity to elucidate the study by creating a storyline with rests of the categories. The core analytical category of this study shows increased post-disaster vulnerabilities of female-headed households due to unequal gender relations.

Results

The concept of gender describes the socially approved attributes, role, activities and responsibilities that are connected to an individual through their sex. These are socially constructed rather than biologically determined. It determines that being a male or female, how individuals are perceived in society and role expectations in the given social structure. There are set of social relationships between both males and females, known as gender relations that are basically concerned with distribution of power between both the sexes. These are social inequalities that can be produced and reproduced through social institutions, norms and practices. Thus, unequal gender relations are blamed for creating and reproducing systemic differences between men and women's position in society. Gender relations are not uniformed, and there are variations according to class, race, ethnicity, disability, differences of time and place, and over the time, they might change. Many societies try to establish them as natural way of organizing roles and relation between men and women. In much of the South Asian countries, including India, women are given lower social status than men. They are confined to household responsibilities, without financial resources, and completely dependent on their male counterparts.

Institutions and Gender Relations

Social institutions are conceived as norms, values and code of conduct expressed through traditions, cultural practices, customs and formal–informal laws. North (1991) has defined institutions as framework of rules for achieving certain social and economic goals. They shape social and economic opportunities for men and women and decide their autonomy in decision making. In this way, institutions affect development outcome and also produce unequal gender relations. When it comes to social institution and their role in reproducing gender inequalities, North (1991) has referred to specific structural form of institutions that frame meanings and beliefs for the society which affects the individual at larger scale. The need to understand role of social institutions has been emphasized here because gender relations are beyond the boundaries of household. Family and kinship are its primary sites but they are not confined to it. Gender relations are social constructs that can be reproduced (Kabeer and Subrahmanian 1999). To analyse the life experiences of women in the light of gender relations, the study has adopted Naila Kabeer's Social Relations Framework. The framework was developed in early 1990s at the Institute of Development Studies (Sussex University). Through this framework, the study has analysed gender inequalities and discriminatory practices created, maintained and reproduced by four key social institutions: the state, the market, the community/civil society and family/kinship.

Social Relations Framework

The framework has intended an analytical approach to address how social institutions provoke gender inequalities through distribution of resources, responsibilities and power. The larger aim of the framework is to serve as an aid for policies in order to support women as development agents. In a narrow form, the structure can be used in analysing gender relations within single institutions. The study has collectively analysed four institutions, that is, state, market, community and family. State as a larger institution consists of legal, military and administrative organizations, and market ranges from

Table 3.1

Profile of the respondent

S. No.	*Age*	*Education*	*Breadwinner*	*Marital Status*	*Dependents*
1	47	Graduation	Self	Widow	3
2	37	Graduation	Self	Widow	1
3	38	Graduation	Self	Widow	3
4	37	Graduation	Self	Widow	1
5	42	Post-graduation	Self	Widow	2
6	30	Graduation	Self	Never married	2
7	47	Uneducated	Son	Widow	NA
8	47	Primary	Son	Widow	NA
9	44	Primary	Son	Widow	NA

firms, financial corporations, farming enterprises and multinational. The third institution, community, is made up of various supra family groups, village tribunals, political factions, neighbourhood networks and non-governmental organizations. The last institution is family, which consists of household, extended family and lineage groupings as major components.

The profile of participants of this study is presented in Table 3.1

Analysis and Discussion

Kabeer and Subrahmanian (1996) have stated that family relationships and household are primary sites for reproducing unequal gender relations. The study thoroughly looked into gender relations within the households through certain norms and practices that reproduce unequal gender relations. There are certain rules imposed on women attributing to their social status and role expectations which vary through different stages of their lives.

Family Institutions and Gender Inequalities

The data reveals that rigid gender norms are followed within the households which have larger social implications on gender relations. During crisis situations, family emerged as a primary support, however, further probe unfolded layers of family dynamics and gender discriminatory practices. The following excerpts from respondents' interviews throw light on intra-household dynamics and establish the role of family practices in promoting vulnerabilities:

> My brother completed his education and he was already working but I was forced to get married at an early age. The school was far off so girls in our village were not allowed to go that far. I used to help my mother with domestic work. Rather than my studies, my parents were concerned about my marriage and at the age of 18, I got married.

The practice of early marriage was observed amongst respondents with rural background. There are severe implications of getting married at a very early age such as being burdened with household duties, poor reproductive health and having limited opportunities and interaction beyond household. One of the early consequences was to force girl child to leave school so that they can learn household responsibilities.

> I was studying in class 11th when I got engaged and after that I left school for a year. After my engagement, my family insisted me to learn household work. I would help my mother with domestic chores. I rejoined school for my intermediate examinations only. Later, I had to leave studies due to my marriage. My younger sister also got engaged when she was studying in school.

The gender bias among these households was clearly visible through level of education among male and female child. A respondent informed that although her brothers were elder to her however, she was the first one to get married while her brothers pursued their career and got married later than her.

> My parents were highly concerned about my marriage and by the time they couldn't find a groom for me, I was allowed to continue my studies.

Adding further to this, another respondent's experience depicts gender inequality within the household.

> I never went to a school or Madrasa. My brothers went to schools and as they grew up they owned family business, my mother used to teach me household skills.

Pertaining to gender discrimination in education, another respondent shared the following:

> I had to leave school after primary because after primary there were no schools for girls in my village. My brothers went to nearby village to attend school but I stayed back home to help my mother with household work and learn cooking, sewing and other such skills.

Contrary to this, few of the respondents informed that pertaining to education their families were very encouraging. As a respondent informed that she had two more sisters but her parents supported all of them to complete their education and become financially independent. She was working before her marriage and after her marriage also she continued her work. Similarly, more participants informed how their families encouraged them to complete education and become financially independent; however, a significant factor was poor financial status of these households where income sources were unsustainable. Hence, the women had to come out of the household at a very early age to contribute to the family income. A respondent informed,

> My father died when I was 8 years old. Due to lack of education my mother was unable to work and my siblings were younger than me. My mother was completely financially dependent on her brothers. My brother started working after his schooling. I was also working with many NGOs during my college days.

Similarly, another respondent belonging to poor financial background shared:

> My father was the only breadwinner in the family. He was doing petty jobs and often fell ill, so I and my sister started working after completing school education. Our salaries contributed to the medical expenses.

Even after starting to work and having contributed to the family economy, the respondents informed that their role as breadwinner of the family is not as recognized as that of a male breadwinner. For instance, a participant revealed the following:

> My in-laws kept insisting me to leave my job. They would argue that I belong to a well to do family so I do not need to work rather I should look after my kids. My father-in-law often used to get angry if I was late from office. My family often complained that due to office, I am not giving time to my children and other household responsibilities.

Another participant informed that her in-laws never appreciated her educational qualification rather her suggestions on important issues were considered as interference in family decisions. Another respondent shared her in-laws' disagreement on her job.

> I worked as a social worker and I would go to slums and talk to people with various background. My in-laws didn't like this and often insisted me to leave my job. My husband also wanted me to do the same ... to understand his family, and agree with them.

The patriarchal norms have always emphasized that ideal women should be limited to the household; however, the repercussions of such practice are never anticipated. From the given excerpts, it was observed that gender discriminatory practices within the households are one of the primary factors in predetermining their vulnerabilities. As the data reveals that women who had completed formal education were able to take up the jobs offered by railways, whereas uneducated or less-educated women were not eligible for the job. They had to transfer these jobs to their relatives. Thus, these vulnerabilities exist throughout their life cycle and in the wake of disasters are revealed and intensified.

Gender Relations within Community

Community is the second institution of this framework which is composed of village tribunals, voluntary associations, informal networks, patron–client relationships and NGOs. Community/ religious practices and beliefs are considered an integral part of the community in strengthening and bringing people closer. They are considered a set of sacred practices that help nurtures communities; however, this category of analysis will look into cultural practices particularly focusing on widowhood. Traditionally in Indian context, widows are considered a marginalized group with a poor social status and social, economic and cultural deprivation. Widows are considered inauspicious for the family and communities and are often forced to move out of family and take refuge in *ashrams*. It is an experience full of hardships for any woman and it's been made miserable by the experience of exclusion and separation from their in-laws and society.

Many of the respondents admitted calling off relationship with their in-laws after their husband's death as they were blamed for bringing bad omen to the family. For example, a respondent remembered how her relationship turned sour after her husband's death.

> Neither could they take (in-laws) anyone's responsibility and nor they wanted to keep any relationship with anyone.

Another respondent also admitted that after her husband's death, in-laws called off relationship with her.

> My in-laws have not communicated with me at all since his death. At present I am not considered as part of my in-laws' family. They blamed me for bringing bad luck to their son.

Another respondent stated,

> My in-laws have called off their relationship with me. Since the day I left their house, there has been no communication between us. After my husband's death they had a strong opinion that my presence is inauspicious for their family. They hated my son and blamed him also for taking their son's life. I did not receive any emotional or financial support from my in-laws.

Another respondent shared the treatment she received from her in-laws after her husband's death.

> My brother-in-law used to talk to me and my kids very rudely. He would taunt me for being a widow and blamed me for his brother's death.

Adding to the similar experience, another respondent admitted the changes that she experienced in the behaviour of people after her husband's death.

> There were very few people who would come to meet me. Close relatives and in-laws were more interested in the compensation money that I had received. They used to ask for their share from that money.

These excerpts depict the stigmas attached to widowhood and also throw light on changes in family behaviour after a woman becomes widow. They are excluded from the family by the mere thought that a widow is inauspicious for the family. Such practices increase hardships for women because earlier they were completely dependent on their husband and in-laws. As they enter widowhood, they experience sudden withdrawal of support on which they were banking upon for years. This sudden lack of social capital limits their interaction with the public sphere and weakens their support system.

The study also observed instance of gender inequality through cultural practices and rituals. One of the respondents spoke on rituals after her husband's death which revealed the following:

> In Islam, initial days of widowhood are strictly followed by the practice called Iddat, which is a period of 4 months and around 10 days. During this period a widow is not allowed to communicate with any male who does not share any blood relation with her. Thus, only my children, brother and father could have a face to face communication with me.

The deeper meaning of this practice is to avoid any communication with widow that might lead to establishing marital relationship or attraction to the opposite sex. However, the respondent denied practicing *Iddat* due to the sudden burden of household responsibilities. Another respondent shared why she denied the traditional practices of not wearing her *Mangalsutra* which is supposed to be shed off by widows.

> I wear it for my security. I do not want any men to look at me as helpless or lonely woman.

Through these excerpts, it was observed that these practices of eliminating an individual from sociocultural environment contribute towards increased social vulnerabilities. The household financial decisions became challenge for these women as they had been dependent on their husbands so far. The exclusion from immediate family left them even more miserable as they were now more dependent on their extended families. There poor legal understanding was a challenge in claiming the compensation and also most of them recalled struggling with their in-laws to not share the monetary compensation. On the contrary, there are few instances of receiving complete support from the family. The elderly widows had larger support from their kids and extended families. The restrictions on clothing and appearance were not strictly imposed on them as compared to the widows of young age. As a 47-year old respondent said,

> I still wear bright color sarees except Red, I would dress up on occasions and my family doesn't have any issues. It is my own decision to wear what I want.

Market and Gender Relations

In this category, market as a social institutions and its role in reinforcing unequal gender relations has been analysed. Market refers to firms, financial corporations, farming enterprises and multinationals. Feminist theorists have conceived market as a capitalist institution which exploits women by taking their labour for free or underpaid. Participation of women in income generating activities is a source of bringing change in their overall cultural and economic status. However, on analysing the data it revealed that attaining financial independence is not the only source to change the traditional social status of women. The discrimination with male and female breadwinner has been discussed earlier. Here the discrimination in public sphere will be discussed through the following excerpts which reveal instances of underpayment and odd working hours:

> "I had to stay for more than required hours in the lab."
> "The kind of job and salary I was offered was very less as compared to the amount of work I was doing in the office."

> "Moreover I could not tolerate the rude behavior of my seniors. Most of them were not concerned about my problems at workplace."
> "I was taken for domestic help in households of my seniors and these kinds of helps were not remunerative."
> "Rather than my name, I was addressed as a 'safaiwali' which I found derogatory."
> "As compared to the amount of traveling and work that was included in this job, I was not paid enough."
> "I am unsatisfied with my work profile and working time. Since I hold a master's degree, I was expecting a better profile in terms of salary and position, but that did not happen. Also I have to shuffle my timing after every week."
> "From health point of view, late night shifts are not a very ideal time for work at my age. They do not have off on Sundays so on holidays also I am unable to spent time with my children. After night shifts, I am often unable to take proper rest due to household responsibilities. Sometime I have to go out and get some work done due to which I can't take proper rest."
> "The salary paid by railways is so less that there are hardly any savings".

Here, we see that the state policy provided these women an opportunity to come out of their private sphere and interact with the public sphere however; the oppression is still prevalent in the form of underpaid jobs and unsatisfying workplace. The jobs are offered under Category D which was a major reason that most of the families, especially women, denied these jobs. The gender roles imposed on them, availability of financial resources and age were such factors which didn't allow them to move out of the household and work for a low paying job. One of the respondents denied working as a cleaner because she belonged to higher caste and it wouldn't have been considered appropriate within her family.

State and Gender Relations

The last social institution in the framework of social relation analysis is the state which includes military and administrative organization. Over here *state* refers to the government which makes welfare policies. This institution aims at addressing different form of inequalities within the society. It also holds the power and authority to create, reproduce or demolish gender inequalities/equalities through policies. In context of the study, the role of the state is analysed in reproducing gender inequalities as it is the most powerful structural part of the public sphere. As part of the compensation policy, the jobs were offered to the victims' immediate relatives and women were given priority. On one hand, it seems a tool to empower the weaker gender by eradicating the immediate financial burden that women go through after losing breadwinner of the family. On the other hand, there were many flaws in the policy. As part of the compensation policy, women with certain educational qualification could only opt for these jobs; educationally underqualified and illiterate women were deprived of this opportunity and were forced to transfer their job to eligible family member. Second, women were offered underpaid jobs as compared to their educational qualification and financial capacity. The overqualified women were unable to find out suitable employment for themselves due to the pre-existing gender inequalities and gender role expectations at household and community level. Thus, the policy turns out to be ignoring the aspect of gendered needs and capacities. It was commonly assumed by the policy makers that offering jobs in the lowest category would cater to the needs of the households. It is important that the welfare state should look beyond providing the bare minimum. Additionally, in order to remove gender inequality through policies, particularly in disaster management, the focus should have been at providing employment to all women as per their capacities, skills and eligibilities. Thus, by making gender-sensitive policies for welfare of the larger group, gendered vulnerabilities which emerge after such disasters can be reduced.

Conclusion

The study made an attempt to analyse how unequal gender relations establish gendered vulnerabilities through layers of institutions. These vulnerabilities are identified at various incidents; however, disasters and emergency situations are social catalysts which not only affect the infrastructural development but also tremble the social structure as they largely affect its resources. When disasters threaten, women have lesser resources to escape because of predefined unequal gender relations. When there is scarcity of resources, the already discriminated population, precisely women, are hit even harder. The arguments presented in the study prove that female headship as a single factor can't be condemned for post-disaster increased vulnerabilities of women. Rather, it's a reflection of sociocultural patterns of gender relations with social institutions and household boundaries. Changes in the existing social structure take place over a long span and also do not confirm permanent role reversal. The gendered vulnerabilities are poorly understood but it is apparent that they are deeply rooted in nexus of unequal gender relations. Therefore, the study urges on the formulation of more gender-sensitive relief and compensation policies. The social changes are subject to various contributing factors; however, gender-sensitive disaster management policies will certainly contribute towards lesser gendered vulnerabilities.

Notes

1. http://www.rediff.com/news/2006/jul/11train.htm (Accessed on January 5, 2017).
2. Right to Information Act 2005 mandates timely response to citizen's requests for government information. It is an initiative taken by Department of Personnel and Training, Ministry of Personnel, Public Grievances and Pensions, to provide an RTI Portal Gateway to the citizens for quick search of information on the details of first Appellate Authorities, PIOs and so on amongst others, besides access to RTI-related information/disclosures published on the web by various public authorities under the Government of India as well as the state governments. http://rti.gov.in/ (Accessed on May 2, 2017).

References

Ariyabandu Malalgoda Madhavi and Wickramasinghe, Maithree (2005). *Gender Dimensions in Disaster Management: A Guide for South Asia*. Sri Lanka: ITDG Practical Action South Asia Publication.

Drabeck, Thomas E. and McEntire, David A. (2003). "Emergent Phenomenon and the Sociology of Disasters, Lessons, Trends and Opportunities from the Research Literature", *Disaster Prevention and Management: An International Journal*, Vol. 2, No. 2, pp. 97–112.

Enarson, Elaine (1998). "Through Women's Eyes: A Gendered Research Agenda for Disaster Social Sciences", *Disasters*, Vol. 22, No. 2, pp. 157–173.

Enarson and Phillips, (2008). "*Invitation to a New Feminist Disaster Sociology: Integrating Feminist Theory and Methods*", in Phillips Brenda and Morrow Betty (eds). *Women and Disasters: From Theory to Practice*. Philadelphia, Xlibris Publications, pp. 41–74.

Flick, Uwe (2009) *An Introduction to Qualitative Research*. New Delhi: Sage Publications.

Fothergill, Alice (1996). "Gender, Risk, and Disaster", *International Journal of Mass Emergencies and Disasters*, Vol. 14, No. 1, pp. 33–56.

Fothergill, Alice (2004). *Heads Above Water: Gender, Class, and Family in the Grand Forks Flood*. Albany: State University of New York.

Fothergill, Alice and Peek, Lori (2004). "Poverty and Disasters in the United States: A Review of Recent Sociological Findings", *Natural Hazards*, Vol. 32, pp. 89–110.

Fordham, Maureen (1999). "The Intersection of Gender and Social Class in Disasters: Balancing Resilience and Vulnerability", *International Journal of Mass Emergencies and Disasters*, Vol. 17, No. 1, pp. 15–36.

Ikeda, Keiko (1995). "Gender Differences in Human Loss and Vulnerability in Natural Disasters: A Case Study from Bangladesh", *Indian Journal of Gender Studies*, Vol. 2, No. 2, pp. 171–193.

Kabeer, Naila and Subrahmanian, Ramya (1996). "Institutions, Relations and Outcomes: Frameworks and Tools for Gender Aware Planning". Available at https://www.ids.ac.uk/files/Dp357.pdf (Accessed on May 2, 2017).

Kabeer, Naila and Subrahmanian, Ramya (1999). *Institutions, Relations and Outcomes: A Framework and Case Studies for Gender Aware Planning*. New Delhi: Kali for Women.

Krishnaraj, Maithreyi (1997). "Gender Issues in Disaster Management: The Latur Earthquake", *Gender, Technology and Development*, Vol. 1, No. 3, pp. 395–411.

Mehta, Manjari (2007). *Gender Matters: Lessons for Disaster Risk Reduction in South Asia*. Kathmandu: International Centre for Integrated Mountain Development.

Moghadam, Valentine M. (2005). "The Feminization of Poverty and Women's Human Rights", SHS Papers in Women's Studies/Gender Research, No. 2. Available at http://www.cpahq.org/cpahq/cpadocs/Feminization_of_Poverty.pdf (Accessed on May 2, 2017).

Morrow, Betty Hearn and Enarson, Elaine (1996). "Hurricane Andrew through Women's Eyes: Issues and Recommendations", *International Journal of Mass Emergencies and Disasters*, Vol. 14, No. 1, pp. 5–22.

Neumayer, Eric and Plümper, Thomas (2008). "The Gendered Nature of Natural Disasters: The Impact of Catastrophic Events on the Gender Gap in Life Expectancy", *Annals of the Association of American Geographers*, Vol. 97, No. 3, pp. 551–566.

North, Douglass C. (1991). "Institutions", *The Journal of Economic Perspectives*, Vol. 5, No. 1, pp. 97–112.

Pincha, Chaman (2008). *Indian Ocean Tsunami through the Gender Lens: Insights from Tamil Nadu*. India: Oxfam America /NANBAN Trust.

Pearce, Diana (1978). "The Feminization of Poverty: Women, Work and Welfare", *The Urban and Social Change Review* (Special Issue on Women and Work), Vol. 11, pp. 28–36.

Quarantelli, Enrico L., Rodriguez, Havidan, and Dynes, Russell (2007). *Handbook of Disaster Research*. New York: Springer Publications.

Rose, Gillian (1993). *Feminism and Geography: The Limits of Geographical Knowledge*. Cambridge: Polity Press.

Standing Committee on Railways (2011). *New Railway Recruitment Policy*, Ministry of Railways, Lok Sabha. New Delhi: Government Press.

Strauss, Anselm and Corbin, Juliet (1998). *Basics of Qualitative Research: Techniques and Procedures for Developing Grounded Theory*. New Delhi: Sage Publications.

Walby, Sylvia (1990). *Theorizing Patriarchy*. Cambridge: Basil Blackwell Publications.

Wough, William L., Jr., and Sylves, Richard T. (2002). "Organizing the War on Terrorism", *Public Administration Review*, Vol. 62, pp. 145–153.

World Health Organization (2002). "Gender and Health in Disasters" Available at http://apps.who.int/iris/bitstream/10665/68886/1/a85575.pdf (Accessed on May 2, 2017).

4

Gender Difference in PTSD, Mental Health and Quality of Life

A Focus on Female Tsunami Survivors

Neethu P.S. and Surendra Kumar Sia

Introduction

India is generally marked by regular terror attacks, natural disasters, poverty, riot and ineffective healthcare systems. These conditions often boost the risk of exposure to potential traumatic events. The tsunami on 26 December 2004 severely affected the Union Territory of Puducherry and the coastal regions of the Tamil Nadu and Kerala. Natural disasters such as tsunami are traumatic event, which causes disturbances in the physical, social and mental health of the victims (Lindemann 1994). Victims of such environmental disturbances are affected by trauma-related psychological disturbances and are more prone to develop posttraumatic stress disorder (PTSD; Norris et al. 2002).

PTSD is an anxiety disorder which occurs in people who have witnessed or experienced any traumatic or life-threatening events such as serious accidents, natural disasters, war or rape (National Centre for PTSD 2010). PTSD became the focus of attention since 1980 when it was first incorporated in the diagnostic nomenclature of *DSM-III* by the American Psychiatric Association (APA). In 2013, the fifth edition of *Diagnostic and Statistical Manual of Mental Disorders* (*DSM-5*) came out with a revised PTSD diagnostic criterion.

In which, the first and most criteria to meet PTSD are to experience a traumatic event. The Group A of criterion is related to the stressors, which specify that the person should have direct or indirect exposure or have witnessed any traumatic event including death, threatened death, actual or threatened sexual violence or serious injuries. The intrusion symptoms such as intrusive memories, traumatic nightmares, dissociative reactions, marked physiological reactivity and intense or prolonged distress after exposure to traumatic events were included in Group B criterion. The Group C criterion includes persistent effortful avoidance of trauma-related thoughts, feelings and external reminders. Criterion D specifies the negative alterations in cognition and mood, whereas criterion E implies alterations in arousal and reactivity. The Groups F, G and H signify the persistence of symptoms, functional significance and exclusion criteria (American Psychiatric Association 2013). Usually, the symptoms of PTSD may disappear with time in some cases, but in some individuals it may persist for years. In addition, these symptoms often contribute to depression, anxiety, memory problem, substance abuse and other mental and physical health issues (Brown and Wolfe 1994; McFarlane and Papay 1992). Only fewer studies were conducted in Indian context on PTSD. As such, studies on the intervening role of constructs are lacking. So in the present study, we are analysing the impact of PTSD on the quality of life of the tsunami survivors as well as the mediating role of mental health.

PTSD, Mental Health and Quality of Life

The impact of PTSD reflects on all the functional areas of the victims. Earlier literatures revealed that the subsistence of PTSD may lead to family and marital adjustment problem (Jordan 1992), violent behaviours (Chemtob et al. 1994), occupational problems (Zatzick et al. 1997), low performance at education (Stein et al. 2000) and poorer role functioning (Jordan 1992). Literatures also highlighted that PTSD impairs the vitality, physical and mental health (Schonfeld et al. 1997). Higher rates of depression, substance abuse, dissociation and suicides were found among individuals with PTSD (Briere and Runtz 1988; Hendin and Hass 1991). These problems in turn curtail the quality of life of the victims. Angermeyer and Kilian (1997) defined quality of life as the aspects of life which makes the life fulfilling and worthwhile that extends beyond the anxiety symptoms and include individuals life satisfaction and subjective well-being. A study conducted by Johnsen et al. (2002) revealed that people who experience multiple traumatic events exhibit poor quality of life. Similarly many of the previous researchers also found out that PTSD is associated with lower quality of life among victims of any natural disasters or other trauma-related events (Chou et al. 2004; Heo et al. 2008; Krakow et al. 2002; Miller et al. 2004; Neethu and Rafeeque 2015; Tsai et al. 2007; Warshaw et al. 1993; Zatzick et al. 1997; Zoellner et al. 2000).

PTSD has a greater risk of developing a number of mental health disorders such as anxiety, depression, substance abuse and eating disorders (Brewerton 2007; Harned 2006; Scott et al. 2008). Individuals with PTSD have about five times more chance to develop anxiety disorder and about six times more chance for depression than people without PTSD (Green and Kimerling 2004). Researches from the last few decades were focused on the impact of psychiatric disorders on the quality of life of the individuals. Empirical studies revealed that anxiety disorders hamper the physical and mental health, family and social relationship as well as the home and work functioning of the patients (DuPont et al. 1996). Accordingly, anxiety disorder has a significant negative effect on quality of life (Gladis et al. 1999; Mendlowicz and Stein 2000). Whereas the impact of anxiety disorder on quality of life is independent of demographic variables, diagnostic comorbidity and symptom severity (Cramer et al. 2005; Rapaport et al. 2005; Strine et al. 2005). In a similar vein, a few researchers suggested that if PTSD is prolonged and severe, it leads to depression (Goenjian et al. 1995; Najarian et al. 1996; Terr et al. 1999). Moreover, depression is a strong predictor of quality of life (Tsay and Healstead 2002). Depressive symptoms show a stronger negative relationship with quality of life (Killingworth and Akker 1996; Steele et al. 1996). Rather than this, PTSD results in psychiatric numbing and occupational and recreational impairment which in turn hinder the interpersonal relationships (Solomon 1989). So, social dysfunction is an important outcome of catastrophic events such as natural disasters (Erikson 1976; Titchener and Kapp 1976). Overall researches suggest that PTSDs have been associated with poor quality of life (Chou et al. 2004; Heo et al. 2008; Krakow et al. 2002; Miller et al. 2004; Zoellner et al. 2000) and mental health (Brown and Wolfe 1994; McFarlane and Papay 1992).

Gender Difference

There are some pre-existing socially based gender differences in response to trauma. Basically, gender is associated with how we perceive and expect to think and act as men and women based on the way the society is organized, and not on the basis of our biological differences (World Health Organization 1998). Aggression for men and anxiety or depressions for women are tending to be the most acceptable ways of response to distress or trauma, which results in different patterns of posttraumatic symptoms (Tolin and Breslau 2007). Men and women respond differently to traumatic events based on their immediate traumatic cognitions and behaviours. Compared to males, female victims show more self-blame for the event and hold greater belief that the world is dangerous and they are incompetent and damaged (Tolin and Foa 2002). Several demographic factors influence the development of PTSD.

A study conducted by Giaconia et al. (1995) emphasized that family environment has a great influence on the development of PTSD. Similarly, Ditlevsen and Elklit (2012) revealed the influence of demographic factors such as gender and age on the development of PTSD. In a similar vein, Kessler et al. (1995) found that gender is a great risk factor for the development of PTSD, and they also added that females are more vulnerable to PTSD than males (Breslau et al. 1999; Stein et al. 1997, 2000). Women meet the criteria for PTSD about double than men (Breslau et al. 1998; Kessler et al. 1995) and are four times more likely to develop chronic PTSD (Breslau and Davis 1992). The PTSD is more likely to develop in females than males after exposure to traumatic events, and this difference is greater if the exposure occurred in childhood (Breslau et al. 1997, 1998).

Researches pointed out that there are some gender stereotypes in mental health problems, that is, women are prone to more emotional problems, whereas men are prone to alcohol problems (Afifi 2007). Women reports more worry than men (Robichaud et al. 2003). Overlooking into the gender difference on particular mental health problems, it was noticed that depression and anxiety disorders are more likely to diagnose in women than in men (Parker and Hadzi-Pavlovic 2004; Silverstein 2002; Simonds and Whiffoen 2003). The reason for this existing gender difference is because women tend to have more negative orientation, so they engage in more thought suppression (Robichaud et al. 2003). Somatic depression was found to be higher among women than men (Silverstein 2002). Depression is always comorbid with anxiety. Simonds and Whiffoen (2003) found that either anxiety or depression alone or comorbidity was diagnosed in women than in men. So the ratio of either anxiety alone or in combination with depression is higher among women than men.

Based on the empirical evidences we formulated our hypothesis as:

H1: Posttraumatic stress disorder will have negative correlation with quality of life among tsunami victims.

H2: Posttraumatic stress disorder will have positive correlation with somatic complaint, anxiety, social dysfunction and severe depression among tsunami victims.

H3: There will have significant gender difference in PTSD, mental health and quality of life among tsunami victims.

H4: Somatic complaints, anxiety, social dysfunction and severe depression will sequentially mediate the relationship between PTSD and quality of life among female tsunami victims.

Figure 4.1

Proposed serial mediation model

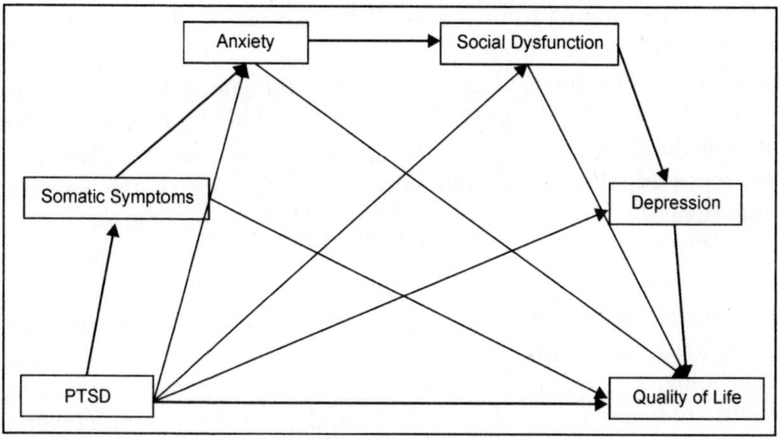

Method

Participants

Participants include 235 tsunami victims, comprising 73 males and 162 females who were selected randomly from tsunami-affected areas of Pondicherry, India. The tsunami on 26 December 2004 caused significant damage to Pondicherry, about 599 deaths were reported and 10,061 dwelling units were lost or damaged (Government of Puducherry, 2005). The Pondicherry Government rehabilitated all the survivors into a newly built settlement named 'tsunami quarters'. We have collected the data from the victims living in the tsunami quarters. Initially 250 questionnaires were distributed among the participants; out of this, 235 completed questionnaires were collected back. All the participants were within the age group of 30–45 years, with an average age of 35.5 years. Participants who experienced death or other traumas in the family for the last one year were excluded from the study.

Tools

The following three self-reported measures were used:

The Posttraumatic Disorder Checklist: The Posttraumatic checklist (PCL) was developed by Weathers et al. (2013). PCL has three versions, and the present study adopted the PCL-C version for collecting data from the tsunami victims. This is an easily administered self-report rating scale comprising 17 items, which assess the *DSM-IV* symptoms of PTSD. Responses were rated on a five-point Likert scale ranging from 'not at all' to 'extremely', thus the possible total scores ranges from 17 to 85. It has excellent psychometric properties. The internal consistency recommended by the authors ranges between 0.94 and 0.97. Test–retest reliability is 0.96. Sample item includes 'In the past month, how much were you been bothered by repeated, disturbing, and unwanted memories of the stressful experience?'.

WHO Quality of Life-BREF (WHOQOL-BREF): World Health Organization Quality of Life Brief Version (1998) is a 26-item self-report scale, which is an abbreviated version of the WHOQOL-100. It contains four dimensions such as physical health, psychological health, social, environmental and global estimate. Items are scored on a five-point scale, and some items require reverse coding. Internal consistencies and test–retest coefficients were above 0.70 for all domains. Sample item includes, 'To what extent do you feel your life to be meaningful?'.

General Health Questionnaire (GHQ-28): The General Health Questionnaire was developed by Goldberg and Hillier (1979) for screening the emotional distress and psychiatric morbidity. It is a

28-item screening tool consisting of four dimensions namely: somatic symptoms (includes items 1–7), social dysfunction (items 15–21), anxiety or insomnia (items 8–14) and severe depression (items 22–28). Sample item includes, 'Have you been getting scared or panic for no good reason?'. Items were scored in a four-point Likert scale ranging from 'not at all' to 'much more than usual'. The test–retest scores was 0.85.

Results

Table 4.1 presents the descriptive statistics and intercorrelation analysis of the study constructs. The results of the independent sample *t*-test were presented in Table 4.2. Statistical Package for Social Science (SPSS) version 20 was used for analysing the data. Then we tested a five-path mediation effect. PROCESS macro using SPSS v20 was used to test the mediation effect (Preacher and Hayes 2004).

Table 4.1

Descriptive statistics and correlations among study variables

Variables	*Mean*	*SD*	*1*	*2*	*3*	*4*	*5*	*6*	*7*
Gender	1.43	0.50	1						
PTSD	37.54	23.14	0.95**	1					
QOL	29.66	7.54	–0.95**	–0.90**	1				
SS	7.69	1.02	0.56**	0.50**	–0.46**	1			
Anxiety	16.06	9.20	0.84**	0.85**	–0.77**	0.57**	1		
SDy	13.34	7.97	0.87**	0.90**	–0.78**	0.56**	0.96**	1	
Depression	11.69	5.16	0.92**	0.79**	–0.89**	0.63**	0.73**	0.72**	1

$N = 235$

$^{**}p < 0.01$ level of significance.

PTSD: Posttraumatic stress disorder, QOL: Quality of life, SC: Somatic symptoms, SDy: Social dysfunction.

Gender is significantly and positively correlated with PTSD (r = 0.95), somatic symptoms (r = 0.56), anxiety (r = 0.84), social dysfunction (r = 0.87) and depression (r = 0.92), whereas significantly and negatively correlated with quality of life (r = –0.95). Table 4.1 also depicts that the independent variable PTSD is showing significant negative correlation with the dependent variable quality of life, the correlation coefficient is –0.95. Result also shows that all the mental health variables (somatic symptoms, anxiety, social dysfunction and depression) are significantly and positively correlated with PTSD, and significantly and negatively correlated with quality of life.

Table 4.2

Gender difference in the study variables

	Gender	*N*	*Mean*	*SD*	*t (sig)*
PTSD	Male	73	18.70	1.76	–24.37***
	Female	162	62.67	11.26	
QOL	Male	73	35.80	2.43	24.53***
	Female	162	21.47	2.40	
SC	Male	73	7.20	0.41	–5.52***
	Female	162	8.30	1.21	
Anxiety	Male	73	9.40	1.93	–12.87***
	Female	162	24.93	7.32	
SDy	Male	73	7.40	.67	–14.33***
	Female	162	21.27	6.09	
Depression	Male	73	7.60	.81	–14.33***

$N = 235$

***$p < 0.00$ level of significance.

PTSD: Posttraumatic stress disorder, QOL: Quality of life, SC: Somatic symptoms, SDy: Social dysfunction.

Table 4.2 presents the gender difference in PTSD, quality of life and mental health dimensions (somatic symptoms, anxiety, social dysfunction and severe depression). The result shows that the mean value of PTSD for males is 46.80 and for females is 48.66, and the *t*-ratio is 3.85 ($p = 0.001$) which indicates significant difference between males and females on the experience of PTSD. That is females are experiencing more PTSD than males. The result also indicates significant difference in quality of life among males and females ($t = 2.22$). Males are showing more scores on quality of life (98.15) than females (96.07). Males are also reporting high mean values on mental health (67.01) than females (65.39), the *t*-value is 2.92 which is significant at 0.05 level. So the results of *t*-test portrayed that there is significant gender difference in PTSD, quality of life and mental health among tsunami survivors. Females are experiencing more PTSD, less quality of life and mental health.

The direct and indirect effect of the structural model is presented in Table 4.3. Figure 4.2 also represents the estimates of the structural model. To test the mediation effect, bias corrected and

Table 4.3

Path coefficients and indirect effects for mediation model

	Path Coefficients					*Indirect Effects*	
	To SS	*To An*	*To SDy*	*To SD*	*To QOL*	*Estimates*	*Bias-Corrected Bootstrap 95% Confidence Interval*
PTSD	.10	.24	.15	.13	-.28		
SS		.55	.29	.78	.81		
Anxiety			.59	.11	–.05		
Sdy				.23	.19		
SD					–.36		
PTSD→SS→QOL						0.09	–0.01, 0.19
PTSD→An→QOL						–0.01	–0.08, 0.04
PTSD→SDy→QOL						0.03	–0.03, 0.08
PTSD→SD→QOL						**0.05**	**0.03, 0.11**
PTSD→SS→An→QOL						–0.01	–0.07, 0.02
PTSD→SS→SDy→QOL						–0.001	–0.02, 0.001
PTSD→SS→SD→QOL						0.04	–0.03, 0.03
PTSD→An→SDy→QOL						0.03	–0.01, 0.10
PTSD→An→SD→QOL						**–0.01**	**–0.05, –0.04**
PTSD→SDy→SD→QOL						**0.01**	**0.01, 0.05**
PTSD→SS→An→SDy→QOL						–0.02	–0.03, 0.02
PTSD→SS→An→SD→QOL						0.01	–0.001, 0.11
PTSD→SS→SDy→SD→QOL						–0.03	–0.01, 0.00
PTSD→An→SDy→SD→QOL						**0.01**	**0.01, 0.05**
PTSD→SS→An→SDy→SD→QOL						–0.01	–0.01, 0.00

percentile bootstrap methods were conducted in PROCESS macro using SPSS v20. In the present study, we have four single mediation models. Model one is the mediating role of somatic complaints on the relationship between PTSD and quality of life; model two is the mediation role of anxiety. Models three and four are the mediating role of depression and social dysfunction on the relationship between PTSD and quality of life respectively. According to the serial mediation notion, we integrated all these four single mediation models and formed a five-path serial mediation model. Direct effect and indirect effect of PTSD, somatic symptoms, anxiety, social dysfunction and depression on quality of life were also checked. The path coefficient should be more than 0.1 to influence the endogenous variable (Urbach and Ahlemann 2010), and there should not be any zero in between the upper and lower limit of the bias corrected bootstrap confidence interval for the significant paths. The result represents that the direct effect of PTSD on quality of life is –0.28. The indirect paths show that severe depression is fully mediating the relationship between PTSD and quality of life (0.05) and it exhibits a suppression effect. Anxiety and severe depression (–0.01) is sequentially mediating the exogenous–endogenous relationship and finally anxiety, social dysfunction and severe depression (–0.001) were fully and sequentially mediating the effect of exogenous variable PTSD on the endogenous variable quality of life.

Figure 4.2

Final serial mediation model

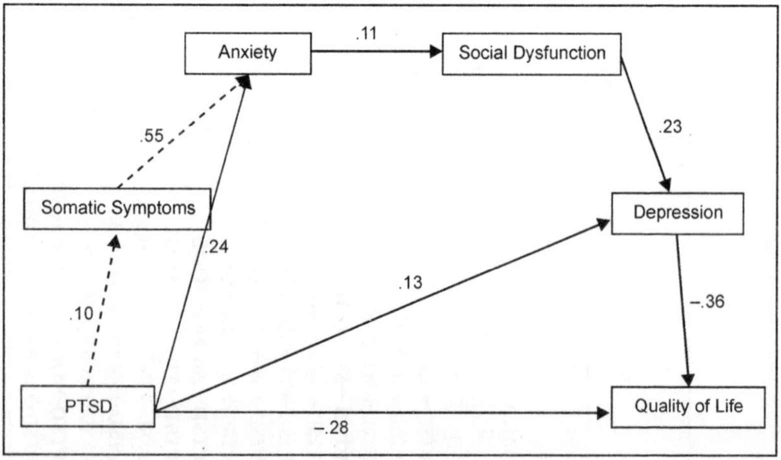

Discussion

Natural disasters such as tsunami cause various symptoms of PTSD and other mental health problems among the victims. Similarly the present study also analysed the PTSD, quality of life and mental health among tsunami survivors. The findings of the correlation analysis were supporting with the first hypothesis that 'PTSD will have significant negative correlation with quality of life among tsunami survivors'. Result shows that PTSD is negatively correlated with quality of life. The finding is supporting with the previous study finding by Johnsen et al. (2002), he revealed that people who experience multiple traumatic events exhibit poor quality of life. Similarly many of the other researchers also found out that PTSD associated with lower quality of life among victims of any natural disasters or other trauma-related events (Chou et al. 2004; Heo et al. 2008; Krakow et al. 2002; Miller et al. 2004; Neethu and Rafeeque 2015; Tsai et al. 2007; Warshaw et al. 1993; Zatzick et al. 1997; Zoellner et al. 2000). Victims of any traumatic events have more chances for meeting the criteria of posttraumatic disorders. The symptoms of PTSD often co-occur with other mental and physical health problems, which hinder the quality of life of the victims.

Natural disasters such as tsunami have a negative impact on individuals' mental health. Not only does PTSD curtail the quality of life of the individuals, but it also leads to mental ailments such as depression, anxiety, memory problem, substance abuse and other mental health issues (Brown and Wolfe 1994; McFarlane and Papay 1992). Similarly the results were supporting with our second hypothesis that is PTSD will have positive correlation with mental health dimensions such as somatic symptoms, anxiety, social dysfunction and depression.

The second objective of the study was to find out the gender difference in PTSD, quality of life and mental health among tsunami survivors. The findings indicate that women are showing higher rates of PTSD symptoms than men. Previous studies also indicate that women meet the criteria of PTSD double than that of men (Breslau et al. 1998; Kessler et al. 1995) and are four times more likely to develop chronic PTSD (Breslau and Davis 1992). This gender difference is primarily due to the pre-existing socially influenced gender differences in response to traumas. According to our society and culture, women show depression or anxiety and men show aggression in response to any trauma. Moreover female victims show more self-blame for the event and hold greater belief that the world is dangerous and they are incompetent and damaged (Tolin and Foa 2002). This will worsen the severity of trauma, so women are more prone to get PTSD. As previous researchers proved PTSD is linked with mental health and quality of life. The present study findings also indicate that as women are experiencing more PTSD symptoms than males, their quality of life and mental health are also lower than that of males. Higher rate of PTSD hindered the quality of life as well as mental health of the women victims. The result indicates that women are exhibiting more anxiety, somatic symptoms, social dysfunction and depression as well as lowered quality of life than men in the present study. The findings are corroborating with the previous researches that women are more prone to depression and anxiety than men (Afifi 2007), because women tend to have more negative orientation, so they engage in more thought suppression (Robichaud et al. 2003). Considering the circumstances of life in India, women having higher prevalence of life experiences such as sexual assault and rape (Kessler et al. 1995) and other negative life experiences, this might be a contributor to the existing gender differences in PTSD. As girls and women experience different kinds of traumatic events, the prevalence of PTSD and mental health problems increases than men.

As significant gender difference was found in PTSD, quality of life and dimensions of mental health such as somatic symptoms, anxiety, depression and social dysfunctions are gone for a mediation analysis by focusing only the female tsunami survivors. A five-path serial mediation model with the mediating role of mental health dimensions on the effect of PTSD on quality of life was checked. The result shows that instead of a five-path serial mediation model, a four-path model is significant in the present sample. That is the direct effect of the exogenous variable PTSD on the endogenous variable quality of life is sequentially mediating by anxiety, depression and social dysfunction. The findings from the present study are an addition to the existing literatures. As far as the researchers' knowledge, no attempts have been made by any researchers on the mediation effect of mental health on PTSD and quality of life relationship, except few studies on the direct effect of the constructs. Green and Kimerling (2004) revealed that individuals with PTSD are about five times more chance to develop anxiety disorder and about six times more chance for depression than people without PTSD. Similarly, few of the previous literatures also show that PTSD has a greater risk of developing a number of mental health disorders such as anxiety, depression (Brewerton 2007; Harned et al. 2006; Scott et al. 2008) and social dysfunction (Erikson 1976; Titchener and Kapp 1976) and hinder the interpersonal relationships (Solomon 1989). Accordingly, anxiety (Gladis et al. 1999; Mendlowicz and Stein 2000) and depression (Killingworth and Akker 1996; Steele et al. 1996; Tsay and Healstead 2002) have a significant negative effect on quality of life.

Conclusion

Tsunami severely affected overall health of the victims. From the findings, it can be concluded that PTSD had greater influence on the quality of life and mental health of the tsunami victims. Quality of life and PTSD are highly correlated. The samples in the present study victimized the tsunami in 2004, now even after 12 years the PTSD symptoms are still persisting in them, which indicates that the tsunami caused chronic post-traumatic symptoms among these peoples. The mental health and quality of life of the peoples are at risk in these 12 years. Moreover, the findings revealed that women victims are more at the risk of PTSD than males, and their quality of life and mental health were more worsen. Findings also indicate that women are showing more PTSD, somatic complaints, anxiety, depression and social dysfunction and low quality of life than men. Interestingly, the study reveals that anxiety, depression and social dysfunction were sequentially mediating the effect of PTSD on quality of life among female tsunami survivors. Depression is found to be a significant mediator which influences the effect of exogenous variable on the endogenous variable.

By understanding the augmented vulnerability of women to PTSD, government and policy makers can design programmes or trainings for the women in natural calamity-prone areas to prevent the occurrence of PTSD. Women's mental health affects others in society. Counsellors and therapists can initiate appropriate interventions and therapies to reduce the PTSD symptoms and enhance their quality of life and mental health, and can provide more psychological care to the victims. Rescue and rehabilitation workers can also train the women to take necessary precautions.

The present study sheds light into the chronic impact of 2004 tsunami on the life of the victims, even though the study has some limitations. Firstly, the study has taken a very small sample size and preferred a single geographical area, which is not enough to generalize the results. Most of the previous studies were considered only the acute impact of the disaster on the physical and mental health of the victims, but the present study analysed the chronic impact as well. Even though, the study has taken only quality of life and mental health as the study variables; the chronic impacts of tsunami on other aspects of their life such as occupation, education, family and marital adjustment should also be considered in further studies. Further studies can also adopt mixed methodology approach; it will help to probe more data from the survivors. This study adopted only mental health as a mediator, so further studies can be done with inculcating more mediators as well as moderators.

References

Afifi, M. (2007). "Gender Differences in Mental Health", *Singapore Medical Journal*, Vol. 48, No. 5, p. 385.

American Psychiatric Association (2013). *Diagnostic and Statistical Manual of Mental Disorders* (5th ed.). Washington, DC: Author.

Angermeyer, M. C. and Kilian, R. (1997). "Theoretical Models of Quality of Life for Mental Disorders", in *Quality of Life in Mental Disorders*. New York: Wiley, pp. 112–117.

Breslau, N., Chilcoat, H. D., Kessler, R. C., Peterson, E. L. and Lucia, V. C. (1999). "Vulnerability to Assaultive Violence: Further Specification of The Sex Difference in Post-Traumatic Stress Disorder", *Psychological Medicine*, Vol. 29, No. 4, pp. 813–821.

Breslau, N. and Davis, G. C. (1992). "Posttraumatic Stress Disorder in an Urban Population of Young Adults: Risk Factors for Chronicity", *The American Journal of Psychiatry*, Vol. 149, No. 5, p. 671.

Breslau, N., Davis, G. C., Andreski, P., Peterson, E. L., and Schultz, L. R. (1997). "Sex Differences in Posttraumatic Stress Disorder", *Archives of General Psychiatry*, Vol. 54, No. 11, pp. 1044–1048.

Breslau, N., Kessler, R. C., Chilcoat, H. D., Schultz, L. R., Davis, G. C., and Andreski, P. (1998). "Trauma and Posttraumatic Stress Disorder in the Community: The 1996 Detroit Area Survey of Trauma", *Archives of General Psychiatry*, Vol. 55, No. 7, pp. 626–632.

Brewerton, T. D. (2007). "Eating Disorders, Trauma, and Co-morbidity: Focus on PTSD", *Journal of Eating Disorders,* Vol. 15, No. 4, pp. 285–304.

Briere, J. and Runtz, M. (1988). "Post Sexual Abuse Trauma: Data and Implications for Clinical Practice", *Journal of Interpersonal Violence*, Vol. 2, pp. 367–379.

Brown, P. J. and Wolfe, J. (1994). "Substance Abuse and Post-traumatic Stress Disorder Comorbidity", *Drug and Alcohol Dependence*, Vol. 35, No. 1, pp. 51–59.

Chemtob, C. M., Hamada, R. S., Roitblat, H. L., and Muraoka, M. Y. (1994). "Anger, Impulsivity, and Anger Control in Combat-related Posttraumatic Stress Disorder", *Journal of Consulting and Clinical Psychology*, Vol. 62, No. 4, p. 827.

Chou, F. H., Chou, P., and Su, T. T. (2004). "Quality of Life and Related Risk Factors in a Taiwanese Village Population 21 Months after an Earthquake", *Australian and New Zealand Journal Psychiatry*, Vol. 38, pp. 358–364.

Cramer, J. A., Brandenburg, N., and Xu, X. (2005). "Differentiating Anxiety and Depression Symptoms in Patients with Partial Epilepsy", *Epilepsy & Behavior*, Vol. 6, No. 4, pp. 563–569.

Ditlevsen, D. N. and Elklit, A. (2012). "Gender, Trauma Type, and PTSD Prevalence: A Re-analysis of 18 Nordic Convenience Samples", *Annals of General Psychiatry*, Vol. 11, No. 1, p. 26.

DuPont, R. L., Rice, D. P., Miller, L. S., Shiraki, S. S., Rowland, C. R., and Harwood, H. J. (1996). "Economic Costs of Anxiety Disorders", *Anxiety*, Vol. 2, pp. 167–172.

Erikson, K. (1976). "Loss of Community at Buffalo Creek", *American Journal of Psychiatry*, Vol. 133, pp. 302–305.

Giaconia, R. M., Reinherz, H. Z., Silverman, A. B., Pakiz, B., Frost, A. K., and Cohen, E. (1995). "Traumas and Posttraumatic Stress Disorder in a Community Population of Older Adolescents", *Journal of the American Academy of Child and Adolescent Psychiatry*, Vol. 34, No. 10, pp. 1369–1380.

Gladis, M. M., Gosch, E. A., Dishuk, N. M., and Crits-Christoph, P. (1999). "Quality of Life: Expanding the Scope of Clinical Significance", *Journal of Consulting and Clinical Psychology*, Vol. 67, No. 3, p. 320.

Goenjian, A., Pynoos, R., Steinberg, A., Najarian, L., Asarnow, J., Karayan, I., Ghurabi, M., and Fairbanks, L. (1995). "Psychiatric Comorbidity in Children after the 1988 Earthquake in Armenia", *Journal of the American Academy of Child and Adolescent Psychiatry*, Vol. 34, pp. 1174–1184.

Goldberg, D. P. and Hillier, V. F. (1979). "A Scaled Version of the General Health Questionnaire", *Psychological Medicine*, Vol. 9, No. 1, pp. 139–145.

Government of Pondicherry, Office of the Secretary Revenue and Collector (2005). "Damages Caused by Tsunami on 26-12-2004. Report in Respect of UT of Pondicherry".

Green, B. L. and Kimerling, R. (2004). "Trauma, Posttraumatic Stress Disorder, and Health Status", in *Trauma and Health: Physical Health Consequences of Exposure to Extreme Stress*, ed. P. P. Schnurr and B. L. Green, editors. Washington, DC: American Psychological Association. pp. 13–42.

Harned, M. S., Najavits, L. M., and Weiss, R. D. (2006). "Self-harm and Suicidal Behavior in Women with Comorbid PTSD and Substance Dependence", *American Journal on Addictions*, Vol. 15, pp. 392–395.

Hendin, H. and Haas, A. P. (1991). "Suicide and Guilt as Manifestations of PTSD in Vietnam Combat Veterans", *The American Journal of Psychiatry*, Vol. 148, No. 5, p. 586.

Heo, J. H., Kim, M. H., Koh, S. B., Noh, S., Park, J. H., Ahn, J. S., Park, K.-C., et al. (2008). "A Prospective Study on Changes in Health Status Following Flood Disaster", *Psychiatry Investigation*, Vol. 5, pp.186–192.

Johnsen, B. H., Eid, J., Laberg, J. C., and Thayer, J. F. (2002). "The Effect of Sensitization and Coping Style on Post-traumatic Stress Symptoms and Quality of Life: Two Longitudinal Studies", *Scandinavian Journal of Psychology*, Vol. 43, No. 2, pp. 181–188.

Jordan, B. K. (1992). "Problems in Families of Male Vietnam Veterans with Posttraumatic Stress Disorder", *Journal of Consulting and Clinical Psychology*, Vol. 60, No. 6, pp. 916–926.

Kessler, R. C., Sonnega, A., Bromet, E., Hughes, M., and Nelson, C. B. (1995). "Posttraumatic Stress Disorder in the National Comorbidity Survey", *Archives of General Psychiatry*, Vol. 52, No. 12, pp. 1048–1060.

Killingworth, A. and Van Den Akker, O. (1996). "The Quality of Life of Renal Dialysis Patients: Trying to Find the Missing Measurement", *International Journal of Nursing Studies*, Vol. 33, No. 1, pp. 107–120.

Krakow, B., Schrader, R., Tandberg, D., Hollifield, M., Koss, M. P., Yau, C. L., and Cheng, D. T. (2002). "Nightmare Frequency in Sexual Assault Survivors with PTSD", *Journal of Anxiety Disorders*, Vol. 16, No. 2, pp. 175–190.

Lindemann, E. (1994). "Symptomatology and Management of Acute Grief", *American Journal of Psychiatry*, Vol. 101, No. 2, pp. 141–148.

McFarlane, A. C. and Papay, P. (1992). "Multiple Diagnoses in Posttraumatic Stress Disorder in the Victims of a Natural Disaster", *Journal of Nervous and Mental Disorder*, Vol. 180, pp. 498–504.

Mendlowicz, M. V. and Stein, M. B. (2000). "Quality of Life in Individuals with Anxiety Disorder", *American Journal of Psychiatry*, Vol. 157, pp. 669–682.

Miller, M. W., Kaloupek, D. G., Dillon, A. L., and Keane, T. M. (2004). "Externalizing and Internalizing Subtypes of Combat-related PTSD: A Replication and Extension Using the PSY-5 Scales", *Journal of Abnormal Psychology*, Vol. 113, No. 4, p. 636.

Najarian, L., Goenjian, A., Pelcovitz, D., Mandel, F., and Najarian, B. (1996). "Relocation after a Disaster: Posttraumatic Stress Disorder in Armenia after the Earthquake", *Journal of the American Academy of Child and Adolescent Psychiatry*, Vol. 35, pp. 374–383.

National Centre for PTSD (2010). "Advancing Science and Promoting Understanding of Traumatic Stress". U.S. Department of Veterans Affairs. Available at www.ptsd.va.gov.

Neethu, P. S. and Rafeeque, T. C. (2015). "Impact of PTSD on Quality of Life and Subjective Well-being of Peoples in Tsunami Affected Area at Pondicherry: A Comparative Study", *International Journal of Indian Psychology*, Vol. 2, No. 3, pp. 125–130.

Norris, F. H, Friedman, M. J., Watson, P. J., Byrne, C. M., Diaz, E., and Kaniasty, K. (2002). "An Empirical Review of the Empirical Literature, 1981–2001", *Psychiatry*, Vol. 65, pp. 207–239.

Parker, G. and Hadzi-Pavlovic, D. (2004). "Is the Female Preponderance in Major Depression Secondary to a Gender Difference in Specific Anxiety Disorders?", *Psychology Medicine*, Vol. 34, pp. 461–470.

Preacher, K. J. and Hayes, A. F. (2004). "SPSS and SAS Procedures for Estimating Indirect Effects in Simple mediation models", *Behavior Research Methods, Instruments, & Computers*, Vol. 36, No. 4, pp. 717–731.

Rapaport, M. H., Clary, C., Fayyad, R., and Endicott, J. (2005). "Quality-of-Life Impairment in Depressive and Anxiety Disorders", *American Journal of Psychiatry*, Vol. 162, No. 6, pp. 1171–1178.

Robichaud, M., Dugas, M. J., and Conway, M. (2003). "Gender Differences in Worry and Associated Cognitive Behavioral Variables", *Journal of Anxiety Disorder*, Vol. 17, pp. 501–516.

Schonfeld, W. H., Verboncoeur, C. J., Fifer, S. K., Lipschutz, R. C., Lubeck, D. P., and Bueschiing, D. P. (1997). "The Functioning and Well-being of Patients with Unrecognized Anxiety Disorders and Major Depressive Disorder", *Journal of Affective Disorders*, Vol. 43, No. 2, pp. 105–119.

Scott, K. M., McGee, M. A., Wells, J. E., and Browne, M. A. O. (2008). "Obesity and Mental Disorders in the Adult General Population", *Journal of Psychosomatic Research*, Vol. 64, No. 1, pp. 97–105.

Silverstein, B. (2002). "Gender Differences in the Prevalence of Somatic Versus Pure Depression: A Replication", *American Journal of Psychiatry*, Vol. 159, pp. 1051–1052.

Simonds, V. M. and Whiffoen, V. E. (2003). "Are Gender Differences in Depression Explained by Gender Differences in Co-morbid Anxiety?", *Journal of Affective Disorder*, Vol. 77, pp. 197–202.

Solomon, Z. (1989). "PTSD and Social Functioning", *Social Psychiatry and Psychiatric Epidemiology*, Vol. 24, No. 3, pp. 127–133.

Steele, T. E., Baltimore, D., Finkelstein, S. H., Juergensen, P., Kliger, A. S., and Finkelstein, F. O. (1996). "Quality of Life in Peritoneal Dialysis Patients", *The Journal of Nervous and Mental Disease*, Vol. 184, No. 6, pp. 368–374.

Stein, M. B., Walker, J. R., and Forde, D. R. (2000). "Gender Differences in Susceptibility to Posttraumatic Stress Disorder", *Behaviour Research and Therapy*, Vol. 38, No. 6, pp. 619–628.

Stein, M. B., Walker, J. R., Hazen, A. L., and Forde, D. R. (1997). "Full and Partial Posttraumatic Stress Disorder: Findings from a Community Survey", *American Journal of Psychiatry*, Vol. 154, No. 8, pp. 1114–1119.

Strine, T. W., Kobau, R., Chapman, D. P., Thurman, D. J., Price, P., and Balluz, L. S. (2005). "Psychological Distress, Comorbidities, and Health Behaviors among US Adults with Seizures: Results from the 2002 National Health Interview Survey", *Epilepsia*, Vol. 46, No. 7, pp. 1133–1139.

Terr, L., Bloch, D., Beat, M., Shi, H., Reinhardt, J., and Metayer, S. (1999). "Children's Symptoms in the Wake of Challenger: A Field Study of Distant-Traumatic Effects and an Outline of Related Conditions", *American Journal of Psychiatry*, Vol. 156, pp. 1536–1544.

Titchener, J. L. and Kapp, F. (1976). "Family and Character Change at Buffalo Creek", *American Journal of Psychiatry*, Vol. 133, pp. 295–299.

Tolin, D. F. and Breslau, N. (2007). "Sex Differences in Risk of PTSD", *PTSD Research Quarterly*, Vol. 18, No. 2, pp. 1–8.

Tolin, D. F. and Foa, E. B. (2002). "Gender and PTSD: A Cognitive Model", in *Gender and PTSD*, ed. R. Kimerling, P. Ouimette, and J. Wolfe. New York: Guilford Press. pp. 76–97.

Tsai, K. Y., Chou, P., and Chou, F. H. (2007). "Three-Year Follow-up Study of the Relationship between Posttraumatic Stress Symptoms and Quality of Life among Earthquake Survivors", *Journal of Psychiatry Research*, Vol. 41, pp. 90–96.

Tsay, S. L. and Healstead, M. (2002). "Self-care Self-efficacy, Depression, and Quality of Life among Patients Receiving Hemodialysis in Taiwan", *International Journal of Nursing Studies*, Vol. 39, No. 3, pp. 245–251.

Urbach, N. and Ahlemann, F. (2010). "Structural Equation Modeling in Information Systems Research Using Partial Least Squares", *JITTA: Journal of Information Technology Theory and Application*, Vol. 11, No. 2, p. 5.

Warshaw, M. G., Fierman, E., Pratt, L., Hunt, M., Yonkers, K. A., Massion, A. O., and Keller, M. B. (1993). "Quality of Life and Dissociation in Anxiety Disorder Patients with Histories of Trauma or PTSD", *American Journal of Psychiatry*, Vol. 150, pp. 1512–1512.

Weathers F. W., Litz, B. T., Keane, T. M., Palmieri, P. A., Marx, B. P., and Schnurr, P. P. (2013). "The PTSD Checklist for DSM-5 (PCL-5)". National Center for PTSD. Available at www.ptsd.va.gov.

World Health Organization (1998). *Gender and Health*. Technical paper. Geneva: WHO. Publication reference no.WHO/FRH/ WHO/98.16.

Zatzick, D. F., Weiss, D. S., Marmar, C. R., Metzler, T. J., Wells, K., Golding, J. M., Stewart, A., Schlenger, W. E., and Browner, W. S. (1997).

"Posttraumatic Stress Disorder and Functioning and Quality of Life Outcomes in Female Vietnam Veterans", *Military Medicine*, Vol. 162, No. 10, pp. 661–665.

Zoellner, L. A., Goodwin, M. L., and Foa, E. B. (2000). "PTSD Severity and Health Perceptions in Female Victims of Sexual Assault", *Journal of Traumatic Stress*, Vol. 13, No. 4, pp. 635–649.

5

Dealing with Gender Vulnerability through Mediated Communication in the Tamil Nadu Floods, 2015

S. Mahalakshmi and Sunitha Kuppuswamy

Introduction

Disaster

Disaster is a sudden catastrophic natural event that causes great damage to both people and their property. With an instance of disaster occurrence, the normal functioning of a community is disrupted. In such cases, the affected community will require assistance from outside on a massive scale. Disasters result in deterioration of health, environment, quality of life and ecology. Examples of disasters include drought, flood, earthquake, tsunami, heat waves, landslide, storm and wildfire. The damages caused by the natural disasters are increasing like a wild fire. The damages caused by the natural disasters (values scaled to 2014 US $ from 1995 to 2005 in India alone were US $23,907,666, in the following year it jumped to a whopping US $40,683,247.[1] The damages caused by natural disasters in the past two decades have almost been on the verge of getting doubled. The issue of frequent disaster occurrence is bigger in the case of developing nations since they overpower the coping capacity and hamper the development of people. The resources allocated for poverty alleviation and economic development are diverted for disaster relief and restoration (Thomalla et al. 2006).

Flood is an overflow of water in large quantities over what is generally seen as a dry land. Floods result in inundation of areas and submerging of the land. Flood occurs when water overflows in the water bodies such as rivers, lakes and even for that matter water reservoir due to varied reasons. Flood instances are found to be increasing in on a global scale.

Figure 5.1 depicts a histogram which clearly shows the upward trend of flood occurrences in India in the past two decades (1995–2015).

Figure 5.1

Histogram showing the total number of flood occurrences (1995–2015)[2]

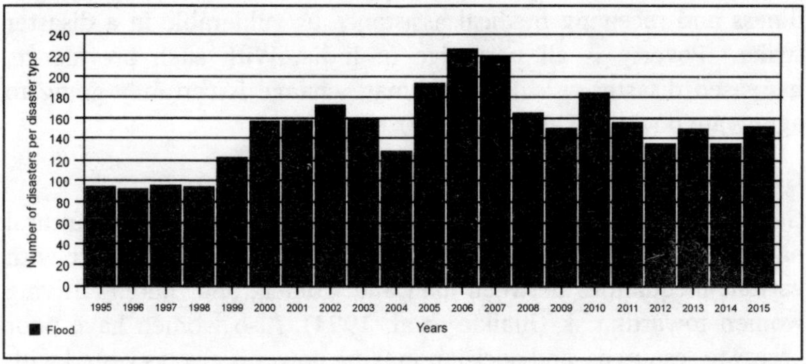

In the year 2015 alone, a total of 152 flood occurrences have been reported of which the Tamil Nadu floods, 2015, was one of the most expensive natural disasters to have occurred in the year.[3] India suffered economy loss amounting to $3 billion due to the Tamil Nadu floods in 2015.

The phases to act upon in a disaster include mitigation, preparedness, response and recovery. In phases of mitigation and preparedness, measures are taken that helps to reduce the impact of hazard by prediction and wherever possible prevent disasters and mitigate their impact on populations considered to be vulnerable to disasters. The response and recovery phases include activities such as providing immediate and long-term assistance to the affected community.

Vulnerability

Vulnerability is defined as the declined capacity of an individual, a community to resist, cope with and recover from a disaster. It implies that a vulnerable person or a community is susceptible, having less exposure to an approaching hazard, poor coping and adaptive capacity due to reasons of physical weaknesses such as poor health and immunity, mental weaknesses such as disability, institutional weaknesses concerning the occupation and place of work and socio-economic weaknesses with respect to income and education (Cutter et al. 2003). According to Wisner et al. (2004), the risk associated with vulnerable communities is higher than others. The potential risk for those include heightened adversities such as injury, loss of body parts, prolonged illness, malnutrition and death. The World Health Organization cites children, pregnant women, elderly people, malnourished people and people suffering from prolonged illness and receiving medical assistance as vulnerable in a disaster strike.[4] Poverty is all pervasive in India. With such prevalence, increased disaster risk due to climate change is certainly going to aggravate poverty (Lal et al. 2009).

Gender Vulnerability

Gender vulnerability sprouts factors such as historical and cultural patterns of relation in a social setup. These factors intersect with various inequalities between men and women, and thereby driving women towards risk (Blaikie et al. 1994). Also women have poor access to resources, and decision-making power is always vested with the men. In the 1991 cyclone in Bangladesh, women succumbed along with their children, since they had to wait for their spouses to return and take a decision with regard to evacuation (WHO 2002).

According to Ahmad (1994), women are still in most places of the world confined to the household. About two decades back in Scotland, a woman was separated from her marital life partly by her husband who blamed her for the loss of their property due to the flood. Because of the fear of the spouse's displeasure, women tend to be afraid to even evacuate from their disaster-hit homes, thus increasing their vulnerability to disasters culturally. It is evident that disaster strikes different people in different ways. Women themselves belonging to different groups, namely, octogenarians, pregnant, those with disorders and illness undergo disasters differently (Drolet et al. 2015). In the case of developing nations, the effects of disasters on women are greater than and different from those on men (Agarwal 1990; Ali 1987; Begum 1993; Drèze and Sen 1989; Jiggins 1986; Kerner and Cook 1991; Sen 1988, 1990; Rivers 1982; Vaughan 1987; Walker 1994). The risk factor for women in disasters increases by multiple folds because of their poor coping strategies, such as essential life-saving skills, access to resources and cultural factors restricting their mobility.[5]

The intervention of public sector in the United States in the aspect of food, nutrition and supplemental feeding for pregnant women and lactating mothers during natural disasters is a key point since their vulnerability is very high because of a greater risk they pose towards morbidity and mortality involving two lives. The concern is also about the risk of acquiring diseases and disorders that will have an impact on the growth of the foetuses.

Role of Media in Disaster Communication

As mentioned earlier, the loss and damage caused by natural disasters are increasing by leaps and bounds. A dire need has aroused to minimize the catastrophic impacts to both life and property by an effective system. Communication comes to aid, since it is central in the effort of public education, early warning, evacuation and post-disaster relief efforts. According to Stal (2013), disasters can be managed efficiently and effectively with appropriate communication in place. The need to exploit the communication technology advancements towards promotion of regional cooperation for disaster risk reduction (DRR) is evident in the case of the Hindu Kush Himalayan Region, which is prone to instances of flood. The need has already been recognized a decade ago with respect to the exchange of real-time hydrological and meteorological data and information support to support flood forecasting in the Hindu Kush Himalayan region (Ahmad and Ahmed 2003; Chalise and Shrestha 2002; ICIMOD 2003). Media is considered as an excellent channel to spread the awareness of an approaching hazard, sensitize people with live coverage of disaster and disseminate prompt situational inputs, drive people to be proactive during and post-disaster and so on. As the news of 2015 Paris attack broke out, millions of users were informed of the same through social media and in turn people started to inform and warn others of the situation instantaneously.[6] This clearly manifests the role of new media in emergency communication.

New media refers to communication tools that make use of digital technologies such as the Internet, and its content is usually available on demand which can be accessed using a digital device. Social media is one among them. Certain activities of disaster phases such as mitigation and response depend on media, since it aides in bridging the gaps between the needs and the needy. The Ice Bucket Challenge (also known as the ALS Ice Bucket Challenge) was where people posted pictures on the social media showing them drenched in a bucket of ice and water to promote awareness on amyotrophic lateral sclerosis (ALS) disease and encourage donations to research on the same. This activity went viral on the social media. It was the most successful fundraising and awareness raising campaign that generated a whopping $115 million for the ALS association and $700,000 for the local affiliate compared to a mere $2.8 million the organization raised the previous year.[7] The campaign was a forerunner for many other bucket challenges successive to the Ice Bucket did prove to be undoubtedly successful. The role of media is multifaceted and with the emergence of new media the power of it just got amplified.

Social Media

The modern technologies evolved in the mid-1990s led to the development of Internet-based applications known as social media (Lindsay 2011). The term *social media* attributes to the interaction among people in which ideas, information, messages and other contents are shared and exchanged through online networks. Social media allows its user to create a profile and cite their preferences with respect to the users with whom the profile and other details are shared. Users can build a network of connections and stay in touch even with good old friends by joining forums. These forums connect users in the network sharing similar interests. Social media allows people to interact, connect, view, create and share content. Social media is built on Web 2.0 technology allowing two-way communication. The user not only gets information but also interacts by expressing views and opinions. User participation and user-generated content defines the social media. Some popular examples of social media include Facebook, Instagram, Twitter, Tumblr, Pinterest and many more. The ease of access offered by the social media is its biggest advantage. Social media is both mobile and web based and can be accessed anytime and anywhere. The power of social media was glorified during the Arab Spring as it became a revolutionary wave of demonstrations, protests, riots and civil wars in the Arab group of countries representing the heights of controversy over an issue (Alhindi et al. 2012). The social media platform facilitates for communication and interaction on the political, controversial and an array of aspects.

Social Media and Information Dissemination in Disaster Response

The conceptual approaches to the use of social media for emergencies and disasters are twofold. It can be used to disseminate information and receive user feedback through its features and can be systematically used as an emergency management tool (Lindsay 2011). Social media sites are ranked in the fourth position with regard to serve as the most popular source to access emergency information. They are utilized by the individuals and groups to warn others of approaching hazards and risks, communicate safety status and so on (Lindsay 2011).

The mushroomed growth of social media paved way for the common man to be empowered with situational awareness, get to the ground and work for disaster relief, engage in virtual volunteering by sharing the information received with other people in his or her network, send and receive emergency and warning alerts, reach out to a larger section of people, involve in fundraising and resource raising activities, collaborate with fellow people to coordinate relief efforts, and the list is endless. This was evidently seen in the Jammu and Kashmir floods, 2014, where a number of posts were shared in Facebook concerning help requests, availability of assistance, safety updates, helpline numbers of control room, situation updates and so on. People sought information from both disaster response organizations and other fellow citizens.[8] This saw the emergence of digital volunteers in emergency conditions. Digital volunteers are those who shared the information they received in the network, thereby aiding disaster response passively.

Facebook came to the rescue of the recent Nepal earthquake that shook the people in the east with a magnitude of 7.8. The moment we hear of a disaster strike, we are anxious in wanting to know the safety status of near and dear ones residing in the affected area. Within few hours of the Nepal earthquake strike, Facebook came up with a tool named 'Safety-Check' which allowed the user to mark him or her safe, and all the members in the person's friend list will get a notification of the same. The tool also had other options to check the status of friends, family members and other acquaintances in the affected area and provide assistance if required. The tool helped in knowing the safety of people in the disaster struck area. The tool was first deployed majorly in the Nepal earthquake, 2015.

Twitter enables its users to post short messages up to 140 characters called 'Tweets'. Users can read, post and re-post (retweet) the tweets. It has been widely used in disaster response around the world. In the Indian context, it was extensively leveraged for the recent Kashmir floods, 2014, and the Tamil Nadu floods, 2015.

Tamil Nadu Floods, 2015

The Tamil Nadu floods, 2015, started off with rains from 8 November to 14 December. It was a consequence of the very heavy rainfall generated from the annual north-east monsoons and a coincidence with the El Nino, a climate change phenomenon. The floods affected the coastal regions of Tamil Nadu, Andhra Pradesh and the Union Territory, Puducherry. The reasons attributed to the flooding include unregulated spatial planning and hyper-urbanization where constructions took place in the water ways and water bodies, poorly maintained reservoirs that led to over filling beyond capacities and the subsequent overnight water release from reservoirs, ill-maintained drainage systems, choked water flow channels and other climate change factors. The unusual extra rains were predicted by the meteorological department. They expected a 111 per cent of the normal rains from the north-east monsoon.

The city of Chennai alone has been subjected to five instances of major flood occurrences between 1943 and 2005 and still flood preparations were washed away with deregulations in various aspects.[9] The Tamil Nadu floods, 2015, was declared to be an intense natural calamity[10] requiring massive assistance across the country in various forms and claimed 470 lives in various rain-related incidents.[11]

Social Media and the Tamil Nadu Floods, 2015

Inundated roads, submerged subways, in some of the areas, water rose to the second floor of buildings, power was cut, gradually mobile towers also ran out of power. It was a no power and no tower situation in most of the areas. Because of the above-mentioned reasons, many of the citizens were displaced, marooned and denied of their basic needs. Social media served as an excellent platform where volunteers posted with offerings, services, the safety status of the people in flood-hit areas, those seeking help for the aged and even octogenarians and other people who required immediate attention through its various features such as posts, hashtags, tweets and retweets. Citizens apart from the government and other disaster response agencies organized themselves into groups in Twitter and Facebook and worked for the disaster response and recovery by coordinating the efforts across sections.

The value of media's role in natural disasters has now been augmented by the emergence of social media (Plance 2012). Social media has now become the new platform to provide useful insights on the disaster situations in various ways such as visual coverage of the situation on ground, amplifying information by providing various and unfiltered angles on the disaster.

The literature reviewed for the topic of research included research articles published in journals and conferences, theses, reports, news articles and books. From the extensive critical review of literature done on the topic of research, missing deficiencies are taken up for the study. Most research dealt with specific set of people or groups of a particular community or gender. Minimal researches included pregnant women in natural disasters and information dissemination pertaining to them. The plight of pregnant women in natural disasters and how the information was disseminated in the social media is projected for the research study.

The theory of 'diffusion of innovation' is rolled in, as the study deals with the adoption of a technology. The theory explains how ideas are circulated among groups of people. It throws light on the conditions that increase or decrease the likelihood of an innovation being adopted by people of a particular group. In the case of the Tamil Nadu floods, 2015, how people adopted the communication innovation and social media for disaster assistance to the most vulnerable pregnant women is considered.

'Uses and gratification' theory also makes sense in the context of the social media and the Tamil Nadu floods, 2015. The social media has been exploited on unimaginable heights (Whiting and Williams 2013). Beginning with making friends, building networks, connecting with people across the globe, communicating on different subjects, marketing, interacting with varied sets of people, using the cutting edge communication features and using it for emergency situations, the social media serves people's varied and substantial needs.

The study entirely deals with the usefulness of the social media in disseminating information concerning disaster response with special reference to pregnant women. The information dissemination is entirely based on the computer-mediated communication (CMC) approach. The theories of CMC are applied to discuss the social media content.

Objectives of the Chapter

1. To analyse the social media content pertaining to gender-based emergency evacuation with special reference to pregnant women during the Tamil Nadu floods, 2015.
2. To understand the usefulness of social media in disseminating information with special reference to evacuation of pregnant women during the Tamil Nadu floods, 2015.

Methodology

The aim of the chapter is to analyse the content of social media with respect to gender-based emergency evacuation of pregnant women during the Tamil Nadu floods, 2015. A content analysis methodology was adopted. Since the research technique serves for objective, systematic and quantitative description of the content of communication.[12] Content analysis as the name suggests analyses the content of a particular medium by taking into account various aspects of the content. The content of the social media communication with respect to the topic of study serves as the data for testing the hypothesis of the study. The study aims to test the usefulness of the medium in emergency situations for the most vulnerable lot of the Tamil Nadu floods, 2015. The results of the content analysis are stated in numbers, based on which inferences are drawn. Facebook content relating to the assistance seeking and assistance providing with respect to pregnant women during the Tamil Nadu floods, 2015, was analysed. The timeline taken for content analysis begins from 1 to 6 December, 2015, marking the peak period of the Tamil Nadu floods, 2015, requiring massive rescue and relief.

Data Collection

The posts were searched with different permutations and combinations of words including pregnant woman, pregnant women, Chennai floods, 2015, Tamil Nadu floods, lady in labour and other keywords. The search results were limited to pages with a view to focus on how groups disseminated information pertaining to pregnant women evacuation during the Tamil Nadu floods, 2015. The data gathered for analysis is from pages where messages were being posted for public view. A total of eight posts yielded for the study through the search. These posts belong to different pages in the social media, Facebook. Variables set for the study included number of likes, comments, shares and the hash tags used. Following are some of the terms associated with Facebook and mentioned frequently in the study.

Friends List: *Friends list* is a network of people (including friends, colleagues, co-workers, acquaintances, etc.) that is built and maintained in Facebook with whom we can message, post and communicate by various ways and means. The network allows us to even categorize the people in assigned groups.

Posts: *Post* is a feature in Facebook which allows the user to feed the message desired on the wall of the user. Wall is the virtual space in the server dedicated to each user of Facebook. It is also known as a status update. Once posted, a notification will reach all people listed in the Friends list. Along with messages, users can also post photos and videos.

Shares: *Share* is a feature in Facebook that allows the user to re-post a message or information or the link of a web article, photo or video. That will eventually notify all the members, who will be able to view the message and in turn re-post the message resulting in making the information go viral.

Likes: *Like* feature allows the user to express their liking towards a particular content in Facebook using emoticons which will again notify the members in the Friends list and also the user will be regularly kept in touch with every other update in the content.

Comments: While trying to share a post in Facebook, the friends in the list of the user are given access to post their opinion on the shared post which will again be notified to the people in the list.

Timeline: *Timeline* is a section in the Facebook user's account that shows the complete story of posts by the user in the reverse chronological order. These posts can be viewed by scrolling down the timeline.

Analysis and Interpretation

The study draws into consideration on pregnant women and natural disasters and how the information dissemination in the social media served them. When compared to the numbers of posts on disaster response on a general aspect, the number of posts with regard to pregnant women was few. These few posts were picked up by many and gradually started to become viral.

The posts analysed in Table 5.1 are listed in the order of inquiry.

Table 5.1

Analysis of variables in the study

Posts	*Page*	*No. of Likes*	*No. of Comments*	*No. of Shares*	*Hash tags*
Army personnel rescued a newborn along with the mother in rain-hit Chennai	NDTV	13,000	321	10,000	#ChennaiRains
	India Today	20,00	64	3,000	#ChennaiRains
	Dinakaran Daily News	52000	175	55000	#ChennaiRains
Post requesting availability of doctors in Mudichur one of the flood devastated areas in Chennai for a lady in labour	Chennai Memes	714	129	23,000	–
	Chennai (a) Madras	12	2	4	–
	Kaatupoochi	9	4	69	#ChennaiRains #PregnantWoman
	Anna University Important Updates	82	2	37	–
Heart melting story of a Hindu couple and their Muslim saviour. When the impact of floods was so much that even boatmen were reluctant to bring in boats for rescue, this young man swam through the flood water and rescued Chitra, who was in labour pain.	ABP News	24000	122	321	#ChennaiFloods #Humanity#India
	ABP Ananda	84000	405	709	#ChennaiFloods
	Mango News	464	8	50	#ChennaiFloods

Cont'd...

...*Cont'd*

Pregnant woman is aided with airlift and medical care near Guindy	The Milli Gazette	229	2	44	–
Chennai floods: IAF airlifts pregnant woman who later delivers twin girls	DNA India	529	7	32	–
	India TV	31,000	80	1,000	–
Pregnant woman along with her first child airlifted as part of rescue operations by IAF. The rescue team airlifted the woman who was seven months pregnant.	The Logical Indian	17,000	168	19,000	–
	The Quint	453	2	34	–
	India Today	2000	22	147	–
	Story Pick	15000	16	103	–
Post informing the availability of a doctor driving from Bengaluru to Chennai to aid pregnant women in labour with contact details	Chennai Rain Relief 2015—CRR	190	30	475	–
Post informing pregnant women about the availability of medical assistance free of cost	Zee News	26,000	32	191	–

Post 1: *Army personnel rescued a new-born along with mother in rain-hit Chennai (see Figure 5.2).*

Figure 5.2

Column chart: Variables in analysis—Post 1

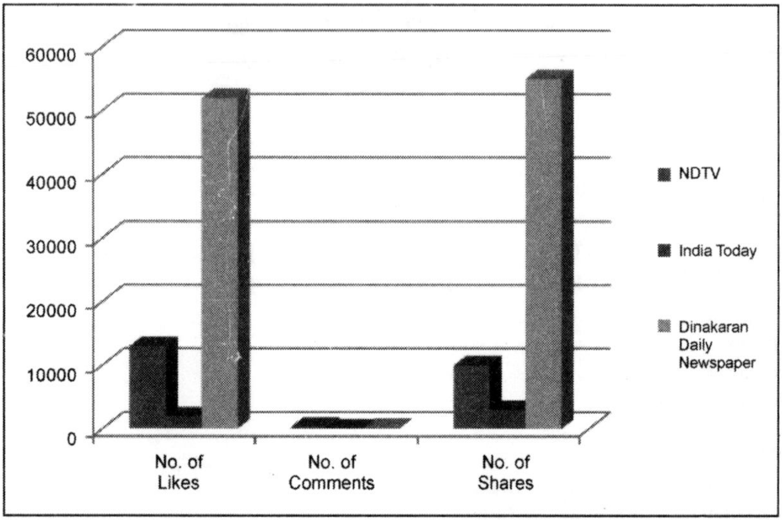

Analysis in point: *Dinakaran* Daily News and NDTV pages were found to be having bigger number of likes and shares. *Dinakaran* Daily News got 52,000 likes and NDTV got 13,000 likes.

Table 5.2

Analysis of the variables—Post 1

Pages	*Likes (%)*	*Comments (%)*	*Shares (%)*
NDTV	19.4	14.7	57.32
India Today	2.99	4.41	11.43
Dinakaran Daily News	77.61	80.88	31.25

More likes paved way for more shares. The same information is posted in all the three pages, yet more attention was drawn to a regional news organization's page (*Dinakaran*) than the national news pages. This indicates that localization of the issue gains predominance in the respective regional media. Comments are seen more on the NDTV page, since the audience belong to an elite educated lot who do not have any hesitance to post their views on the issue. A number of shares are more than the number of likes in the *Dinakaran* Daily News page which shows the social responsibility of their audience in trying to spread the message and reach the intended (see Table 5.2).

Post 2: *Post requesting availability of doctors in Mudichur, one of the flood devastated areas in Chennai for a lady in labour (see Figure 5.3).*

Figure 5.3

Bar chart: Variables in analysis—Post 2

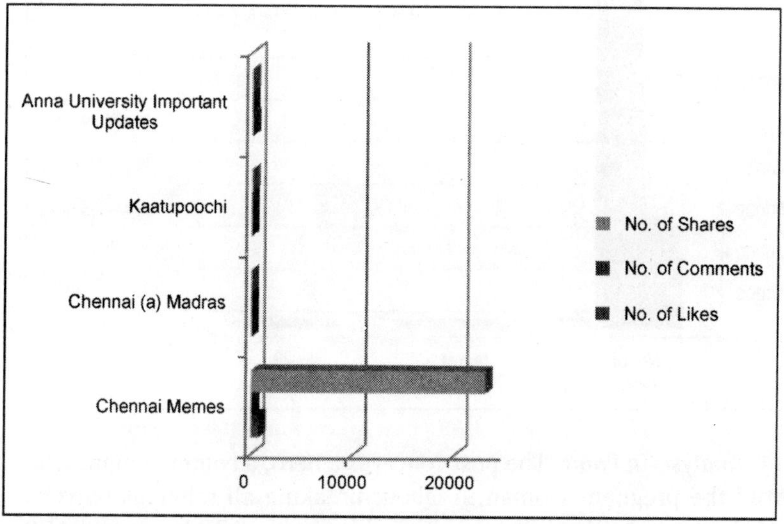

Analysis in Point: The pages containing the posts belong to a mixture of backgrounds, of which the *Chennai Memes* page was found to have the maximum numbers with respect to likes, comments and shares (see Table 5.3).

Table 5.3

Analysis of Variables—Post 2

Pages	*Likes (%)*	*Comments (%)*	*Shares (%)*
Chennai Memes	87.39	94.16	99.52
Chennai (a) Madras	1.47	1.46	0.02
Kaatupoochi	1.1	2.92	0.30
Anna University Important Updates	10.04	1.46	0.16

Post 3: *Heart melting story of a Hindu couple and their Muslim saviour. When the impact of floods was so much that even boatmen were reluctant to bring in boats for rescue, this young man swam through the flood water and rescued Chitra, who was in labour pain (see Figure 5.4).*

Figure 5.4

Column chart: Variables in analysis—Post 3

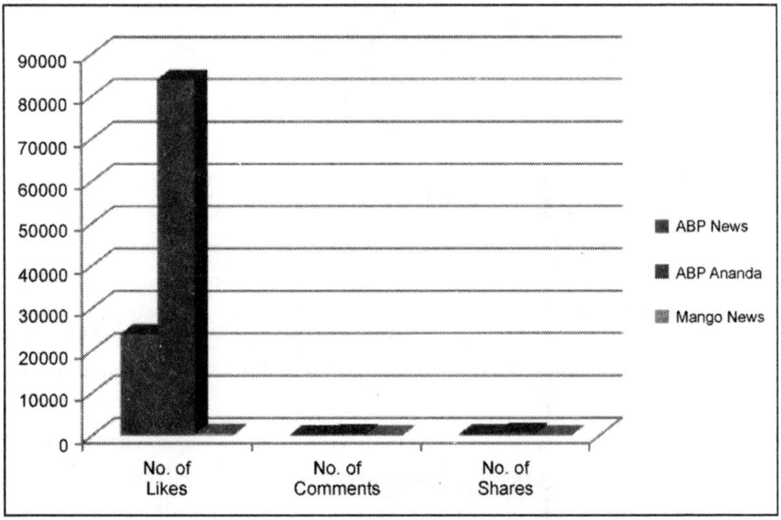

Analysis in Point: The post concerns a hero, a common man, who saved the pregnant woman in labour breaking all religious barriers and the post clearly showcases virtual integrity. ABP News and ABP Ananda belong to the same group. The former is a national news channel page and the latter is a regional language news page. More projection of the news relating to the Hindu pregnant woman saved by a Muslim during Tamil Nadu floods, 2015, projection is seen in the regional media (see Table 5.4).

Table 5.4

Analysis of variables—Post 3

Pages	*Likes (%)*	*Comments (%)*	*Shares (%)*
ABP News	22.13	22.8	29.72
ABP Ananda	77.45	75.7	65.65
Mango News	0.43	1.5	4.63

Post 4: *Pregnant woman is aided with airlift and medical care near Guindy.*

Analysis in Point: There is no scope for comparative analysis in the post mentioned above as the information was posted in only one page (*The Milli Gazette*: a fortnightly published compact English language newspaper).

Post 5: *Chennai Floods: IAF airlifts pregnant woman who later delivers twin girls (see Figure 5.5).*

Figure 5.5

Bar chart: Variables in Analysis—Post 5

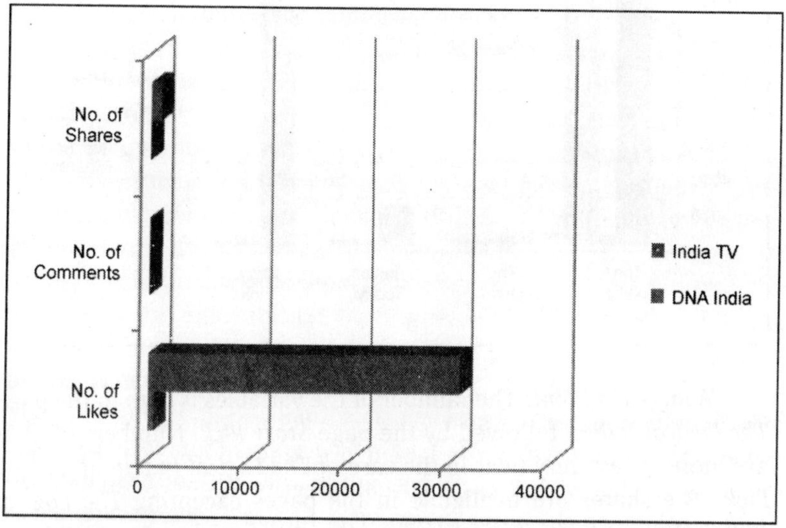

Analysis in Point: The variables considered for analysis is found to be high in the India TV page (see Table 5.5).

Table 5.5

Analysis of variables—Post 5

Pages	*Likes (%)*	*Comments (%)*	*Shares (%)*
DNA India	1.68	8.05	3.1
India TV	99.17	91.95	96.9

Post 6: *Pregnant woman along with her first child airlifted as part of rescue operations by IAF. The rescue team airlifted the woman who was seven months pregnant (see Figure 5.6).*

Figure 5.6

Column Chart: Variables in Analysis—Post 6

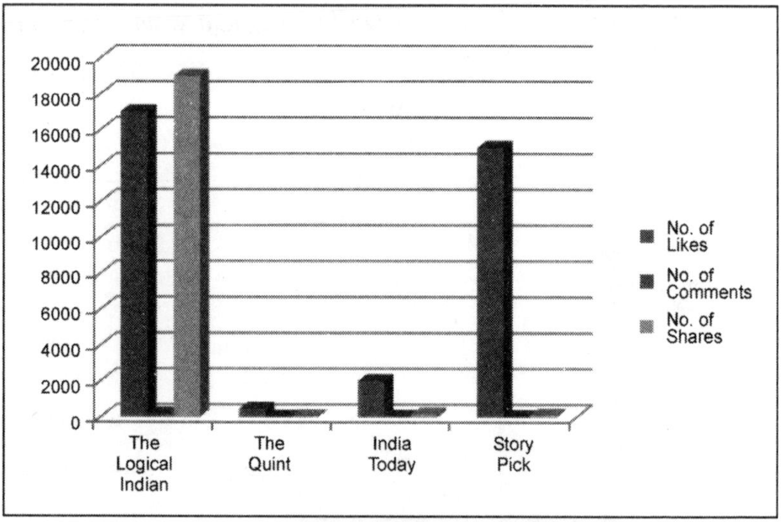

Analysis in Point: The number of the variables is high in the page *The Logical Indian* followed by the page *Story Pick*. Numbers of likes are more or less very near in the case of *The Logical Indian* and *Story Pick*. The shares are negligible in the pages excepting *The Logical Indian*. Users of *The Logical Indian* page are found to be very active particularly with respect to the study. The viral nature of message dissemination is more in *The Logical India* page (see table 5.6).

Table 5.6

Analysis of variables—Post 6

Pages	*Likes (%)*	*Comments (%)*	*Shares (%)*
The Logical Indian	49.34	80.77	98.52
The Quint	1.31	0.96	0.18
India Today	5.81	10.57	0.76
Story Pick	43.54	7.69	0.53

Post 7: *Post informing the availability of a doctor driving from Bengaluru to Chennai to aid pregnant women in labour with contact details.*

Post 8: *Post informing pregnant women about the availability of medical assistance free of cost.*

Analysis in Point: There is no scope for a comparative analysis as both the above-mentioned posts containing information were seen in only one news page.

All of the posts were seen with multiple cues such as text, audio photo and video supporting the media richness theory of CMC. Not all the posts included a hash tag but the ones with hash tag were found with better projection. A total of eight pages were seen with information accompanied by hash tags. The hash tags seen in these posts include: #ChennaiFloods, #ChennaiRains, #PregnantWoman, #Humanity, #India of which the frequency of #ChennaiFloods and #ChennaiRains were more. The correlation between the number of likes and number of shares were not consistent throughout the posts. Not a single post was seen without comments, but with hash tags. As mentioned in the section 'Role of Media in Disaster Communication', millions of people gather to the social media to seek information, the same is evident in the study. Though there were few posts pertaining to evacuation of pregnant women during the Tamil Nadu floods, 2015, the numbers of users who picked them are exorbitant.

Conclusion

Natural disaster is an adverse event that occurs as a result of the natural processes of the earth, causing enormous damage to people and their property. Flood is a natural disaster that occurs when very high volumes of water are displaced due to intense rainfall or other geographic conditions in areas where water cannot be drained. The occurrences of urban flooding are increasing significantly as a consequence of climate change requiring large-scale humanitarian assistance. The role of media, particularly the social media, such as Facebook and Twitter in humanitarian assistance is growing by leaps and bounds. It serves as an excellent platform to request as well as receive humanitarian assistance during a disaster, the recent Tamil Nadu floods, 2015, in particular. The Tamil Nadu floods, 2015, was a recent natural disaster that occurred as a result of very heavy rainfall influenced by the cyclonic circulation over the south west of Bay of Bengal and a strange coincidence with climate change factors. The state's capital and certain coastal regions were severely affected. The disaster claimed more than 470 lives in numerous rain-related incidents.[13]

Gender vulnerability is a major concern in disaster. The elderly women, pregnant women and children are extremely vulnerable because of their coping capacity, susceptibility and exposure. The study aims to examine the 'peer to peer' nature of social media in reducing the gender vulnerabilities during emergency conditions such as the Tamil Nadu floods, 2015. The spirit of solidarity among the humanitarian assistance organizations during the Tamil Nadu floods, 2015, was extensively projected with the help of social media. The projection of information in social media about the Tamil Nadu floods, 2015, in national news pages indicate that a regional issue has gained national focus and drew attention in the national media agenda. Understanding the needs of women is vital in emergency conditions since the disaster recovery groups mostly work through the eyes of men (as seen in the posts analysed). Some of the special needs of women include the requirement of sanitary napkins, counselling services for the distressed after undergoing intense psychosocial trauma due to the disaster and so on.

The analyses clearly show the rationale for special attention to women, identifying their special needs during emergency conditions and address the importance of education, awareness and capacity building in mitigation and preparedness and thereby reducing their gender vulnerability in natural disasters. In the future, social media can be extensively used as a systematic emergency management tool and a medium to plan and roll out campaigns and information dissemination with the intent to improve the awareness levels of women about disasters, the phases and the means and modes of evacuation.

Notes

1. EMDAT (2016). Advanced Search Results of Disaster Statistics. Retrieved from http://www.emdat.be/
2. EMDAT (2016). Disaster Trends. Retrieved from http://www.emdat.be
3. Narasimhan, T. E. (2015). "Chennai Floods Are World's 8th Most Expensive Natural Disaster in 2015". Retrieved from

http://www.business-standard.com/article/current-affairs/chennai-floods-are-world-s-8th-most-expensive-natural-disaster-in-2015-115121100487_1.html

4. WHO Report (2005). "Environment Health in Emergencies". Retrieved from http://www.who.int/environmental_health_emergencies/vulnerable_groups/en/
5. UNEP Report (2011, December 6). "Women Most at Risk from Climate Disasters, Says UN Report". Retrieved from http://populationandsustainability.org/women-most-at-risk-from-climate-disasters-says-un-report/
6. BBC Trending (2015). "How the Paris Attacks Unfolded in Social Media". Retrieved from http://www.bbc.com/news/blogs-trending-34836214
7. McDaniels, A. K. (2015, August 20). "Ice Bucket Challenge More Than a Social Media Phenomenon". *The Baltimore Sun*. Retrieved from http://www.baltimoresun.com/health/bs-hs-als-ice-bucket-20150820-story.html
8. PTI (2014). "J-K Floods: Social Networking Sites Aid Rescue, Relief Work". Retrieved from http://indianexpress.com/article/india/india-others/j-k-floods-social-networking-sites-aid-rescue-relief-work/
9. Sinha, A. and Janardhanan, A. (2015). "Extra Rain and Poor Urban Planning: Why Chennai Went under Water". Retrieved from http://indianexpress.com/article/explained/extra-rain-poor-urban-planning-why-chennai-went-under-water/
10. PTI (2015, December 10). "Centre Has Declared TN Floods as a Calamity of Severe Nature: Jayalalithaa". *The Hindu*. Retrieved from: http://www.thehindu.com/news/national/tamil-nadu/centre-has-declared-tamil-nadu-floods-as-calamity-of-severe-nature-jayalalithaa/article7971800.ece
11. PTI (2016, January 4). "Northeast Monsoon Claimed 470 Lives in Tamil Nadu: Jayalalithaa". *The Hindu Business Line*. Retrieved from http://www.thehindubusinessline.com/news/national/northeast-monsoon-claimed-470-lives-in-tamil-nadu-jayalalithaa/article8064661.ece
12. Berelson, Bernard (1952). Content Analysis in Communication Research. New York, NY: Free Press. 220 pp.
13. PTI (2015). 'TN Floods Claimed 470 Lives, Over 3.83 Lakh Hectare Crops Damaged'. Retrieved from http://www.thehindu.com/news/national/tamil-nadu/tn-floods-claimed-470-lives-over-383-lakh-hectare-crops-damaged/article8064739.ece

References

Agarwal, B. (1990). "Social Security and the Family: Coping with Seasonality and Calamity in Rural India", *Journal of Peasant Studies,* Vol. 17, No. 3, pp. 341–412.

Ahmad, T. (1994). "Women and Water", in *Rivers of Life*, ed. K. Haggard. Dhaka/London: Panos Institute and Bangladesh Centre for Advanced Studies.

Ahmad, Q. K. and Ahmed, A. H. (2003). "Regional Cooperation in Flood Management in the Ganges Brahmaputra Meghna Region: Bangladesh Perspective", *Natural Hazards,* Vol. 28, pp. 191–198. doi:10.1023/A:1021186203100

Alhindi, Waheed Ahmed, Talha, Muhammad, and Ghazali, Bin Sulong (2012). "The Role of Modern Technology in Arab Spring", *Archives des Sciences,* Vol. 65, No. 8, pp. 101–112.

Ali, M. (1987). "Women in Famine", in *Famine as a Geographical Phenomenon*, ed. B. Currey and G. Hugo. Dordrecht: D. Reidel.

Begum, R. (1993). "Women in Environmental Disasters: The 1991 Cyclone in Bangladesh", *Focus on Gender*, Vol. 1, No. 1, pp. 34–39.

Blaikie, P., Cannon, T., Davis, I., and Wisner, B. (1994). *At Risk: Natural Hazards, People's Vulnerability and Disasters*. London: Routledge.

Chalise, S. R. and Shrestha, M. (2002). "Regional Cooperation for Flood Disaster Mitigation in the Hindu Kush Himalayas". In the report of the consultative meeting on developing a framework for flood forecasting in the Hindu Kush Himalayan region, Kathmandu, Nepal, 15–18 May 2001. Kathmandu: ICIMOD.

Cutter, Susan L., Boruff, Bryan J., and Shirley, W. Lynn (2003). "Social Vulnerability to Environmental Hazards", *Social Science Quarterly*, Vol. 84, No. 2, pp. 242–261.

Drèze, J. and Sen, A. (1989). *Hunger and Public Policy*. Oxford: Clarendon Press.

Drolet, Julie, Dominelli, Lena, Alston, Margaret, Ersing, Robin, Mathbor, Golam, and Haorui, Wu (2015). "Women Rebuilding Lives Post-disaster: Innovative Community Practices for Building Resilience and Promoting Sustainable Development", *Gender and Development* Vol. 23, No. 3, pp. 433–448.

ICIMOD (2003). "Regional Cooperation for Flood Disaster Mitigation in the Hindu Kush Himalayan Region": Report of the 2nd high level consultative meeting on establishment of a regional flood information system 10–13 March 2003. Nepal: ICIMOD.

Jiggins, J. (1986). "Women and Seasonality: Coping with Crisis and Calamity", *IDS Bulletin*, Vol. 17, No. 3, pp. 9–18.

Kerner, D. and Cook, K. (1991). "Gender, Hunger and Crisis in Tanzania", in *The Political Economy of African Famine*, R. Downs, D. Kerner and S. Reyna. Philadelphia, PA: Gordon and Breach Science Publishers.

Lal, N. Padma, Singh, Reshika, and Holland, P. (2009). "Relationship between Natural Disasters and Poverty: A Fiji Case Study", SOPAC Miscellaneous Report 678.

Lindsay, R. Bruce (2011). "Social Media and Disasters: Current Uses, Future Options, and Policy Considerations". CRS Report for Congress. Congressional Research Service 7-5700 R41987.

Plance, W. David (2012). "Emergency and Disaster Management". Masters dissertation, American Public University System.

Rivers, J. P. W. (1982). "Women and Children Last: An Essay on Sex Discrimination in Disasters", *Disasters*, Vol. 6, No. 4, pp. 256–267.

Sen, A. (1988). "Family and Food: Sex Bias in Poverty", in *Rural Poverty in South Africa*, ed. T. N. Srinivasan and P. K. Bardham. New York: Columbia University Press.

Sen, A. (1990). "Gender and Cooperative Conflict", in *Persistent Inequalities: Women and World Development*, ed. I. Tinker. Oxford: Oxford University Press.

Stal, Marc (2013). "Disaster and Crisis Communication: Trend Analysis of Technologies and Approaches". Global Assessment Report on Disaster Risk Reduction 2015.

Thomalla, Frank, Downing, Tom, Spanger-Siegfried, Erika, Han, Guoyi, and Rockstorm, Johan (2006). "Reducing Hazard Vulnerability: Towards a Common Approach between Disaster Risk Reduction and Climate Adaptation", *Disasters*, Vol. 30, No. 1, pp. 39–48.

Vaughan, M. (1987). *The Story of an African Famine: Gender and Famine in Twentieth-century Malawi*. Cambridge: Cambridge University Press.

Walker, B. (1994). "Editorial", *Focus on Gender*, Vol. 2, No. 1, pp. 2–6.

Whiting, A. and Williams, L. David (2013). "Why People Use the Social Media: A Uses and Gratification Approach", *Qualitative Market Research: An International Journal*, Vol. 16, No. 4, pp. 362–369.

WHO Report (2002). "Gender and Health in Disasters", Geneva, Switzerland: World Health Organization, Department of Gender and Women's Health.

Wisner, B., Blaikie, P., Cannon, T., and Davis, I. (2004). *At Risk: Natural Hazards, People's Vulnerability and Disasters* (2nd edition). London/New York: Routledge.

PART III

Women's Vulnerability and Responses to Disaster

6

Revisiting Odisha Super Cyclone of 1999

The Barriers and Coping Mechanisms of Disabled Women to Natural Disaster

Pradeep Kumar Parida

> Disability need not be an obstacle to success. I have had motor neurone disease for practically all my adult life. Yet it has not prevented me from having a prominent career in astrophysics and a happy family life.
>
> (Stephen W. Hawking 2011)

> It was a painful and disillusioning realization to recognize that disabled women occupy a multifarious and marginalized position in Indian society, based on their disability and also on sociocultural identities that separate them into categories constructed according to such properties as caste, class, and residential position. Disabled women thus can have plural identity markers that make their daily experience perplexing and difficult.
>
> (Anita Ghai 2002: 49)

In the month of October 1999, within a gap of 10 days, the Indian state of Odisha faced two severe cyclonic storms. The first cyclone from 17 October to 19 October with wind speed of up to 200 kilometre per hour killed 199 people in four districts in southern Odisha (GOI 2002). The second cyclone from 28 October to 30

October was very severe in nature with the wind speed of up to 350 kilometre per hour killed at least 10,000 people and destroyed most of the adjoining coastal areas (IFRC 2000; Oxfam 2000). The second cyclone was identified as a super cyclone by the Indian Meteorological Department (IMD) based on its magnitude and impact (also see, Parida 2008). The super cyclone, followed by heavy flash floods and tidal surge up to 7 metre high, submerged vast areas extending 15 to 30 kilometre inland (OSDMA 2001). Although the state of Odisha has been experiencing the fury of the cyclone since time immemorial in which the coastal hamlets have been worst affected again and again, the super cyclone 1999 in particular was considered as highly disastrous and remarkable due to its sudden havoc as well as long-term differential impact creating a complex crisis in victims' lives.

There is no doubt that the geographic location and climatic condition of Odisha have made it historically prone to multiple hazards such as floods, cyclones and droughts (Bhatta 1997; Government of Odisha 2002; Parida 2002). For instance, excessive rainfall all through the monsoon period causes major floods affecting various people, whereas little rain more often creates drought because of lack of irrigational facilities (Pradhan 2003). However, 'the crucial point about understanding why disasters happen is that it is not only natural events that cause them. They are also the product of social, political and economic environments (as distinct from the natural environment), because of the way these structure the lives of different groups of people' (Wisner et al. 2004: 4). Further, Wisner et al. (2004) rightly noted:

> people live in adverse economic situations that oblige them to inhabit regions and places that are affected by natural hazards, be they the flood plains of rivers, the slopes of volcanoes or earthquake zones. However, there are many other less obvious political and economic factors that underlie the impact of hazards. These involve the manner in which assets, income and access to other resources, such as knowledge and information, are distributed between different social groups, and various forms of discrimination that occur in the allocation of welfare and social protection (including relief and resources for recovery). It is these elements that link our analysis of disasters that are supposedly caused mainly by natural hazards to broader patterns in society. These two aspects—the natural and the social—cannot be separated from each other: to do

> so invites a failure to understand the additional burden of natural hazards, and it is unhelpful in both understanding disasters and doing something to prevent or mitigate them. (p. 5)

The definition of disability is complex, dynamic and very controversial (WHO 2011). Till date, there is no agreement among scholars related to the concrete definition of 'disability'. However, in this chapter, I follow the definition given by International Classification of Functioning, Disability and Health (WHO 2001), which states that disability is a 'dynamic interaction between health conditions and environmental and personal factors'. Wasserman et al. (2016) explain various types of disability such as:

Paraplegia, deafness, blindness, diabetes, autism, epilepsy, depression, and HIV have all been classified as disabilities. The term covers such diverse conditions as the congenital absence or adventitious loss of a limb or a sensory function; progressive neurological conditions like multiple sclerosis; chronic diseases like arteriosclerosis; the inability or limited ability to perform such cognitive functions as remembering faces or calculating sums; and psychiatric disorders like schizophrenia and bipolar disorder (Wasserman et al. 2016).

It is important to note that persons with disabilities are diverse and heterogeneous. They have different personal aspects with variations in gender, age, socioeconomic status, sexuality, ethnicity or cultural heritage. Each has his or her personal preferences and responses to disability (WHO 2011). Moreover, while disability connects with inconvenience, not all individuals with disabilities are equally disadvantaged. Women with disabilities experience the combined disadvantages associated with gender as well as disability, and may be less likely to marry than non-disabled women (Nagata 2003; Rao 2004).

Literature reveals that the study of disability has distinguished between two fundamental approaches such as the 'medical model' and the 'social model' of disability (Emmett and Alant 2006). The medical model talks about disability from the perspective of health condition and the physical inability of the individual to perform a complete action. As a result, it prescribes to follow medical care for individual adjustment and behaviour change. However, the social model explains disability in terms of socially created situations or problems. This model provides additional theoretical tools as 'disability is not seen as an attribute of the individual, but rather as a complex collection of conditions, many of which are created by the social environment. Addressing disability therefore requires social action aimed at integrating people with disabilities into society and modifying the environment to support their full participation in all areas of social life'. (cited in Emmett and Alant 2006: 446, adopted from WHO 2001) Similarly, Albert et al. (2002: 16) provide accounts of the development of the social construct of disability stating that disability, therefore, is a concept distinct from any particular medical condition. It is a social construct that varies across culture and through time, in the same way as, for example, gender, class or caste ... In this sense, disability as a policy issue becomes a cross-cutting social one, rather than something primarily associated with health and individual well-being.

This study employs the 'social model' of disability as it facilitates to explore the social construction of gendered vulnerability. More particularly, it helps to understand at forefront the issues of women with disabilities who have remained unspoken and invisible in their day-to-day life as well as during emergency situations.

The narratives of differential impact of disasters are well established by disaster scholarship (Blaikie et al. 1994; Enarson and Morrow 1997; Wisner et al. 2004); however, the typical issue of disabled women in their daily lives including sudden natural hazards are rarely discussed in disaster research in both the developed and Third World countries. Broadly, disability studies and gender researchers have overlooked multiple issues of women with disabilities in different social context. As Traustadottir (1990: 1) has rightly said, 'almost all research on people with disabilities has assumed the irrelevance of gender as well as other social dimensions such as social class, race, ethnicity, and sexual orientation'. Turning to the disaster literature, surprisingly very less work has been undertaken to recognize and comprehend the specific ways persistent multiple disasters affect disabled women who are considered as one of the more vulnerable elements of communities. More concretely, less information is known regarding the poor disabled women from different social backgrounds and their coping mechanisms during multiple disasters in rural areas. It is useful to mention here that the feminist movement itself has been sluggish in discussing the subject of specific groups of women, including disabled women (Morris 1993). Against this background, this chapter explores and discusses several barriers and the coping mechanism of disabled women in the context of Odisha super cyclone and discovers the ways pre- and post-disaster obstacles including communicational, cultural, economic, environmental, institutional, political, social, attitudinal or structural factors and the cycle of hazardous disasters further hampered and aggravated these women to put them at the lowest level of existence in Odisha.

Methods

Drawing up on a qualitative research from the Jagatsinghpur district of coastal Odisha that experienced the devastating super cyclone in 1999, this chapter attempts to explore the interface between disability and gender, particularly within the context of disaster. A qualitative method is used here to trace various disabled women within the milieu of other social inequalities and disadvantages such as poverty, access to resources, caste and family structure. The compound interconnection such as these has serious connotations not only for families or households in general, but also for women in particular. I interviewed 27 girls and women with disabilities either with congenital or acquired impairments. They were interviewed in different settings in seven villages of the Ersama block of Jagatsinghpur district. Purposive sampling was used for this research. The data was collected through various techniques such as participant observation and in-depth interviews. I have changed the name of the women to protect their confidentiality. The age group interviewed was 18–81 years old disabled women. The respondents selected were those diagnosed with physical disability resulting from injury (e.g., spinal cord injury, amputation), chronic disease (e.g., multiple sclerosis, rheumatoid arthritis) or congenital conditions (e.g., cerebral palsy, muscular dystrophy); sensory impairments consisting of hearing or visual impairments. None had any mental illness and so could respond to verbal communication. All interviews were conducted in Odiya and recorded using a digital audio recorder, except for two interviews due to their refusal. The audio data was translated into English and transcribed by the author. Table 6.1 shows the distribution of disabled women in terms of different types of disabilities in both Odisha and Jagatsinghpur district.

Table 6.1

Distribution of disabled women based on different types of disabilities in Odisha and Jagatsinghpur

State/District	*Residence*	*Total*	*Seeing*	*Speech*	*Hearing*	*Movement*	*Mental*
Odisha	Total	452,421	239,953	31,048	38,414	97,774	45,232
	Rural	392,291	205,024	27,165	34,953	86,501	38,648
	Urban	60,130	34,929	3,883	3,461	11,273	6,584
Jagatsinghpur	Total	11,460	4,905	779	1,137	3,055	1,584
	Rural	10,469	4,307	733	1,058	2,886	1,485
	Urban	991	598	46	79	169	99

Source: Census of India (2001).

Study Area

Odisha, the southeast state of Republic of India, is located between 1749 N and 2234 N latitude and 8127 E and 8729 E longitude. As per Census (2011), it consisted of 30 districts, 223 towns and 51,330 villages. The total population of Odisha is 41,974,218, of which rural population is 34,970,562 and urban population is 7,003,656. The density of population is 270, of which the district of Khurda with 800 people per square kilometre is considered as the most densely inhabited followed by the district of Jagatsinghpur having 682 people per square kilometre (Census of Odisha 2011). Odisha comprises 4.7 per cent of India's landmass and has a 450 kilometre long coastline.

Figure 6.1

State of Odisha and the block map of Jagatsinghpur district

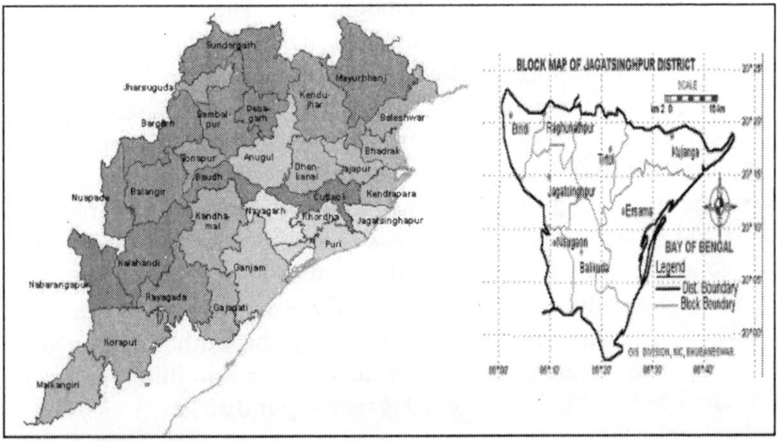

The district of Jagatsinghpur is purposively selected for this study because the impact of the super cyclone was severe in this coastal district. Jagatsinghpur district is situated between 863 E and 8645 E longitude and between 1958 N and 2023 N latitude. Due to the geopolitical location, the district is considered as one of the most natural disaster-prone districts of coastal Odisha. It is encircled by other districts including Kendrapara, Cuttack, Puri and also the Bay of Bengal. The district is also surrounded by the river system of Mahanadi and Paika in north and Devi in the south east. The major economy of the district comes from agriculture and allied products (CDP 2015–16). As per 2011 census, the total population of the district is 1,136,971 including 577,865 male and 559,106 female. Similarly, as per 2011 census, the total SC and ST population is 248,152 and 7,862 respectively. The average literacy rate of Jagatsinghpur district is 86.6 per cent consisting of 92.38 per cent male literacy and 80.6 per cent female literacy (Census of India 2011). There are 8 blocks, 8 tehasils, 1,320 villages and 194 gram panchayats. The name of the blocks and tehasils are Balikuda, Biridi, Ersama, Jagatsinghpur, Kujang, Naugaon, Raghunathpur and Tirtol (CDP 2015–16). My study area falls under the Ersama block. This was purposively selected due to the severe impact of the super cyclone. Seven villages from Ersama block such as Iribina, Japabhuan, Kothi, Singarsahi, Jireilo, Bhiranga and Ersama were purposively selected. The data was collected from November 2016 to March 2017. Table 6.2 provides statistical information about the selected villages, and Table 6.3 presents health, income, education and Human Development Index (HDI) of districts in Odisha.

Table 6.2

Information about the selected villages

Name of the Selected Villages	*Area of Village (Hectares)*	*Number of Households*	*Persons*	*Males*	*Females*
Iribina	255.00	250	958	510	448
Japabhuan	171.00	335	1,358	693	665
Kothi	231.00	200	796	370	426
Singarsahi	303.00	551	2,302	1,146	1,156
Bhiranga	280.00	327	1,251	614	637
Jireilo	218.00	179	767	384	383
Ersama	349.00	527	2,317	1,153	1,164

Source: Census (2011), Jagatsinghpur.

Table 6.3

Health, income, education and Human Development Index of districts in Odisha

District	*Health Index*	*Income Index*	*Education Index*	*HDI Value*	*HDI Rank*
Malkangiri	0.122	0.497	0.491	0.37	**30**
Kandhamal	0.006	0.516	0.645	0.389	**29**
Gajapati	0.173	0.558	0.561	0.431	**28**
Koraput	0.218	0.539	0.535	0.431	**27**
Nabarangpur	0.34	0.453	0.516	0.436	**26**
Rayagada	0.25	0.547	0.531	0.443	**25**
Keonjhar	0.34	0.547	0.704	0.53	**24**
Boudh	0.423	0.497	0.688	0.536	**23**
Jajpur	0.333	0.499	0.786	0.54	**22**
Balangir	0.468	0.504	0.666	0.546	**21**
Ganjam	0.404	0.532	0.718	0.551	**20**
Jagatsinghpur	0.288	0.549	0.833	0.557	**19**
Balasore	0.442	0.466	0.77	0.559	**18**
Bargarh	0.449	0.517	0.727	0.565	**17**
Sonepur	0.474	0.492	0.731	0.566	**16**
Nayagarh	0.462	0.485	0.766	0.571	**15**
Nuapada	0.692	0.47	0.582	0.581	**14**
Sambalpur	0.436	0.59	0.742	0.589	**13**
Dhenkanal	0.468	0.534	0.773	0.591	**12**
Kalahandi	0.763	0.471	0.585	0.606	**11**
Kendrapara	0.596	0.466	0.815	0.626	**10**
Mayurbhanj	0.782	0.489	0.647	0.639	**9**
Bhadrak	0.673	0.463	0.803	0.646	**8**
Puri	0.622	0.527	0.823	0.657	**7**
Angul	0.481	0.748	0.76	0.663	**6**
Deogarh	0.776	0.532	0.698	0.669	**5**
Sundargarh	0.692	0.618	0.74	0.683	**4**
Cuttack	0.686	0.587	0.813	0.695	**3**
Jharsuguda	0.635	0.757	0.773	0.722	**2**
Khurda	0.724	0.639	0.845	0.736	**1**
Orissa	**0.468**	**0.545**	**0.723**	**0.579**	**12**

Source: Planning Commission of India (2004).

Results and Discussion

The Cycle of Disaster: A Gendered Phenomenon

Respondents indicated that they are subject to regular and multiple discriminations in the social, political, familial and economical aspects of their lives. The inferior status of being women with disabilities made them the most disadvantaged group in the respective villages even before the super cyclone. Since most of them were considered as unwanted and unproductive human resources, they received the least attention from various circles including their families. Forming social networks and developing trust was difficult for these women and as a result, they faced multiple problems with regard to disaster preparation. The respondents illustrated how their pre-existing social conditions led them to face the disastrous consequences of the storm. One respondent said,

> When I am considered as an unproductive human resource, it is better not to expect anything from the society or government. (She started crying) I wanted to learn vocational training but I was denied by my family members stating that these trainings are not going to help me in my future.

Majority of the respondents said that their fundamental right including equal inheritance of property was completely denied to them. Some married women with disability expressed doubt whether their husbands would continue to support them when they are not able to perform the duty of wife and mother. From the discussion, a further point that warrants contemplation is that women with disabilities were more vulnerable to harassment and abuse compared to able-bodied women during various cycles of the super cyclone disaster.

Regarding healthcare and rehabilitation facilities, respondents reported that hospitals are mainly located at the district and national level. The SCB Medical College, Cuttack, has the facilities and equipment for critical vulnerable patients, but it was located around 70 kilometre from their place. Respondents also reported that without proper roads in the villages, particularly after a series of floods and cyclones, they found it challenging to reach the SCB Medical College or other superior health centres at Bhubaneswar. So they were dependent on district government healthcare, which usually supplies emergency care and basic needs. Unfortunately, the district government healthcare centres and community level health infrastructures lacked technical capabilities to provide specific care to women with disabilities, reflecting a broader trend in the entire Jagatsinghpur district with regard to the specific vulnerability of the disabled women. Few educated respondents reported that seven women in their villages became victims of polio at different ages due to lack of immunization in their early childhood. The scheme of immunization is still rare in Jagatsinghpur. Had they received treatment on time, most of the polio-affected women would have been saved.

It is discussed in a separate section how disabled women were systematically discriminated against while receiving disaster warning or communication regarding the impending disasters. Usually, disaster communication is a significant support following any disaster, both in the emergency stage and the longer-term recovery stage that follows because it not only assists individuals in saving their lives and property but also facilitates them to cope with the impact. Anita Das, aged 42, from *Dhoba* caste (scheduled caste) with right leg paralysis, narrated her horrific experiences at the time of super cyclone:

> We were five members in our family including two brothers and two sisters. I was the elder one. My father was a farmer and mother a homemaker. Occasionally my mother also worked in our paddy fields. The economic condition of our family was not good, therefore, my parents decided not to send me school although I had interest to study. My father initially tried to send me to school, however, our economic condition did not allow at that time to send four of our brothers and sisters to school at the same time. So my other brothers and sister went to school except me. There was no facility given by the government to help a disabled like me. Since my village is very near to the sea, it was very obvious that any storm would have serious impact on us. There was no proper warning in our village related to the super cyclone; therefore, my father did not show interest to evacuate initially. However, when the storm surge entered into our village, we all wanted to go to a safer place. It was almost one foot height tidal surge initially that made all of us difficult to go to the safer place at the same time. With lot of difficulties, my father was helping each one to go to the safer place, which was around 500 metres from my house. I was the last one to be evacuated. Before that, I tried to crawl in over the dirty floor

> because there was no access for wheelchair. In fact, my disability taught me a great lesson during that emergency situation. My mother died due to snake bite while entering the relatively safer place. From that time till date, my life has been a hell of exploitation. My father got married a second time. It was almost impossible for me to face several challenges, as there was little support from my own family members. I have been a burden in our family. There was no special assistance from the government also. My fear is that if the next disaster with severe magnitude occurs, there will be no one to carry me to the safer places, as I cannot walk due to my cursed and paralysed right leg.

Some upper-caste respondents said that they survived the super cyclone because of their supportive sons and able neighbours in their own area. However, they received little support from the government during the reconstruction period related to intensive support, physical, psychological and otherwise. There was occasional support either from the government or non-governmental organizations. For example, some family members from below poverty line were provided a small concrete house under the scheme Indira Awas Yojana. Conversely, that was not meant to help the women with disabilities. In toto, respondents pointed out that their condition has not changed from the pre-cyclone period till date, and in many cases it has even deteriorated further. The intensely deep-rooted social hierarchies, socio-economic, political and environmental factors as well as their individual impairments have noteworthy consequences for the post-cyclone recovery of women with disabilities.

Incompetent State

As discussed elsewhere, Odisha is a disaster-prone state of multiple varieties. In general, the coping strategy of the people has been weakened owing to the elevated incidence of poverty and unemployment. So, there was no doubt that a high impact disaster like a super cyclone could put severe hurdles in the developmental process. The livelihood security system of the affected areas has been destroyed completely, and there is little scope to come to their earlier positions. However, the question that arises is, were the state and central governments prepared for a calamitous hazard such as a super cyclone? What was the effective post-disaster strategy for the rehabilitation and reconstruction of the vulnerable people such as disabled women? It is noteworthy to point out the responsibility of the government mentioned in Section 8 of the 'Rights of Persons with Disabilities Act, 2016', with regard to humanitarian emergencies and natural disasters:

(1) The persons with disabilities shall have equal protection and safety in situations of risk, armed conflict, humanitarian emergencies and natural disasters.

(2) The National Disaster Management Authority and the State Disaster Management Authority shall take appropriate measures to ensure inclusion of persons with disabilities in its disaster management activities as defined under clause (e) of Section 2 of the Disaster Management Act, 2005, for the safety and protection of persons with disabilities.

(3) The District Disaster Management Authority constituted under Section 25 of the Disaster Management Act, 2005, shall maintain record of details of persons with disabilities in the district and take suitable measures to inform such persons of any situations of risk so as to enhance disaster preparedness.

(4) The authorities engaged in reconstruction activities subsequent to any situation of risk, armed conflict or natural disasters shall undertake such activities, in consultation with the concerned State Commissioner, in accordance with the accessibility requirements of persons with disabilities (The Gazette of India, 2016: 6).

Respondents said that the government of Odisha failed severely on numerous counts. The government's 'soft credit and soft appraisal policy' created a vicious circle of economic vulnerability including food insecurity, nutritional insecurity, education and health insecurity and so on. There was no assistance provided to them in terms of community centres with good living conditions including safety, sanitation, healthcare and counselling. With the recurrence of several consistent disasters in their areas, they faced a serious problem of safe drinking water and appropriate and accessible sanitation facilities. The provisions of comprehensive insurance scheme for women with disability and support for livelihood and upbringing of their children was completely absent. They also did not get caregiver allowance for 'high support' needs. Immediately after the super cyclone, there was no proper mechanism to recognize whether the relief materials reached those who were really affected. Political leaders obtained mileage out of the tragic disaster for the next Assembly and General election taking photos with the victims, and NGOs capitalized plenty of funds collected from different national and international organizations. The situation did not change for several other disasters. Now NGOs are mushrooming, however, not to help the victims but for their own benefit only. In this way, respondents questioned the disaster reduction policy of the Odisha government particularly with reference to women with disability. According to them, disaster reduction policy was made in a 'vacuum without considering the broad socio-cultural, historical, political and ideological framework'. Whatever policy existed, that was not implemented or enforced. For the respondents, an effective policy must involve a process of interaction with members who have different values and priorities. At present, without a specific policy, the lives of women with disability continue to be brutal blueprints of powerlessness and pessimism. In recent years, though several self-help groups have been established in the district, women with disabilities get under-represented in these groups.

In a wider context, however, respondents said that though the 'Rights of Persons with Disabilities Act, 2016', was passed on 28 December 2016, they are still present only on paper. In their villages, there were no measures to ensure that women and children with disabilities enjoy their rights equally with others. Respondents unanimously said:

> We are not provided appropriate environment to utilise our ability and capacity from any governmental sources ... there were no authorities—central, state and local government to ensure that all of us get an equal basis to freely express our views on all matters. We are not provided access to residential and other community support services, including personal assistance. Above all, there was no appropriate mechanism from the government to take measures to protect us from being subjected to torture, cruel, inhuman or degrading treatment.

Generally, for an effective policy implementation, the policy problems are considered in advance of the implementation by the government as it works like a 'central nervous system' (Colebatch 1998). In this connection, Ghai (2002: 50) is sensitive to note about India's legislation for the disabled people:

> Notwithstanding the legislation, the State continues to be largely apathetic. Owing to a complete lack of public planning, responsiveness to any special condition such as disability is missing.

Consequently, the physical environment is largely inaccessible and inconvenient. Dissemination of information in formats accessible to people with sensory impairments is very restricted. For instance, there is a single televised sign-language news bulletin per week for people with hearing impairments. The range of accessible reading materials for people with vision impairments also is extremely limited. The non-availability of other assistive technology devices such as electrically powered wheelchairs at an affordable price adds to the miseries experienced by disabled people.

Of course, the ineffectual nature of the state towards women with disability thus far discussed in this chapter is not by any means to generalize those in terms of good amount of the works already done by the government of Odisha on several occasions after the super cyclone more particularly to reduce the impact of several other disasters. However, the state's initiatives to ensure that women with disabilities enjoy the right to equality, life with dignity and respect for their integrity equally with others is considered to be incompetent and bungling. Therefore, the United Nations Convention on the Rights of Persons with Disabilities can be taken as a mirror regarding the states' responsibilities to people with disabilities during the times of natural disaster:

> States, Parties shall take, in accordance with their obligations under international law, including international humanitarian law and international human rights law, all necessary measures to ensure the protection and safety of persons with disabilities in situations of risk, including situations of armed conflict, humanitarian emergencies and the occurrence of natural disasters. (United Nations 2006, Article 11)

Inaccessible Disaster Communication

Disaster communication is essential for preparing the impending hazards. Shklovski et al. (2010: 1228) say disasters are usually symbolized by 'high levels of information need and low levels of information availability'. For the disabled, communication in advance gives extra timing to negotiate with disaster situations. Respondent narrated their individual problems during the cycle of disasters, For instance, one respondent said, it was difficult to contain flood waters in a wheelchair, if precaution was not taken well in advance. Respondents reported that they failed to receive communication either from the governmental sources or household sources during the super cyclone. Repeated disasters occurred after the super cyclone; however, no change transpired in terms of obtaining effective information and knowledge about the impending disaster. Consequently, the disabled women's initial disadvantage was further heightened by being excluded from the communication channels. Reena Mishra, aged 71 with spinal cord injury, explains:

> People like me face a Herculean task to escape from an impending disaster. Our vulnerability is intensified further when we do not get existing information and facts about the disaster. Unfortunately, in spite of advance warning system, hazards like cyclones and floods are killing people. In my village, I am the witness to uncoordinated relief organisations and inaccurate disaster information in the 1999 super cyclone. In the super cyclone, I was swept away by a powerful tidal surge and fortunately saved by two people in the next village. Nothing was known about my family at that time. I was repeatedly asking information—where is my family, what shall I do now, where do I go after this and so on. These are still unanswered for people like me making little more vulnerable in normal situations. Because of my inadequate physical accessibility and lack of appropriate assistance, I was often deprived of rescue, evacuation services, relief access, safe location, adequate shelter, water and sanitation and other services.

Social Isolation/Exclusion

All the respondents with hearing, lower limb and upper limb disability reported that there was a kind of superstition involved in their respective villages pertaining to their presence in community life. A delusion starts in a family that their attendance can diminish the chances of initiating auspicious work. Subsequently, they are restricted within their parental homes, without being able to perform the traditional role of being a woman. All of them developed feelings of loneliness, self-estrangement and low self-esteem due to their disabilities and social practice. It is not that families are not completely supportive in providing physical assistance to them, but that they frequently fail in delivering emotional support due to their own problems. The fact is that the issues of disabled women are neither discussed in the village community life nor in the village disaster preparedness decision-making meetings to chalk out a future course of action. In fact, there was little scope to identify or recognize this sensitive issue in a larger community life. Even when the super cyclone occurred, disaster managers including municipality or panchayat members were rarely informed about this group who were the most at risk. Though more than 18 years have passed since the super cyclone in Odisha, there is no development in this regard.

More recently, with regard to the gender–caste intersection, Deshpande (2002) advocates that the economic state of women in general continued to be hampered by their caste status in India. The custom of endogamy and other apparatuses directed women's labour and sexuality. He concluded that Dalit women were worse off than upper-caste women in terms of 'standard of living, decision making at a household level, and the prevalence of domestic violence, suggesting that caste, class and gender intersected in increasing or decreasing women's vulnerability'. I found a similar situation with the disabled Dalit women who were guided by the traditional '*varna* system'. In Odisha, although caste practices have undergone changes even in rural areas, ritual restrictions are still prevailing with much force. Respondents from the Dalit castes narrated how caste became more rigid after the super cyclone, although there was a lull during the emergency period. Their ideas corroborate with Gupta (2004: XIII) that 'caste has not changed intrinsically' and reject other viewpoints that caste identities have changed because of independent movement, rapid population growth, education, industrialization, urbanization, the reservation of seats for low castes in the local governing bodies and so on (Government of Orissa 1990; Patnaik 1969). Reiterating similar experiences, Malati Behera, aged 81, from a *Pana jati* (scheduled caste) with vision disability, illustrates:

> My vulnerability does not have any boundary. I am deceived everywhere—by the society, by the government and by the non-governmental organisations. First of all, I was born a disabled woman. That was a punishment meted out to me by *Bhagawan* (God). I see the tragic stories of violence, discrimination and injustice of several women in my villages; therefore, being a woman with disability exacerbates my burden and hardships. Secondly, my *jati* has made me inferior to other women in several contexts. From the pre-super cyclone period till date, I faced discrimination and atrocities. Do you know the pain of being an elderly woman? ... My pain is doubled because I am a disabled elderly woman. Before the super cyclone, my family could not prepare well for the impending disaster. We didn't have the financial resources to go to a new place and settle there. My only son, who is a rickshaw puller, could not

> afford initially to participate in the evacuation process. However, soon the time came when we were compelled to leave our house and settle in the cyclone shelter. I faced tremendous problem in the cyclone centre because nobody was there to listen to my personal agony. There was absolutely no support provided to me even after the cyclone. The relief authority did not visualize my problem. I was not able to stand in the queue and as a result received nothing. My son though he loved me with full devotion, felt frustrated due to my disability status at that time. During the reconstruction period, my specific problem was not addressed by the authority. Even for seeking a bank loan, I had to undergo several constraints. Now, you tell me, am I similar to other people? Is it not the responsibility of the government and other community members to help me during and after the cyclone? Some NGOs approached me and took photos promising that they will provide ₹ 50,000 for my disability, but that did not be materialise. Now, I am waiting to see death.

Further, Mitali Kundu, aged 38, a widow with hearing disability, recited her experience and coping strategy:

> I experienced several discriminations after the death of my husband. The super cyclone was a *kalo* (death trap) for me and my two year old son. The deadly cyclone snatched away my son and mother-in-law from my clutches. I was proved to be an unfortunate woman after that event. After the cyclone, I had to undergo lots of traditional restrictions. I cry every time remembering my husband who was the sole bread winner of my family. However, after his road accident, the entire burden rested on me. My psychological susceptibility stems from various stressors post disaster where I have been humiliated as women headed household in the relief camp and the rehabilitation measures. At this moment, my vulnerability can be related with the loss of life, status in the family, relationships, livelihood and habitation ... You know, I tried to go to my parents' house but in vain as my two brothers did not allow me to stay there for a longer time. Now, I am working as a daily labourer. I am least prepared if another disaster occurs in my area. Unfortunately, I don't get any support from the village community members. Nobody is bothered about my trauma and stress. However, burden of duty and responsibility has heightened my vulnerability to physical, mental and emotional stress. Against all odds, I thank the government of Odisha for providing me a small house under the Indira Awas Yojana.

Insufficient Social Security and Haphazard Relief Distribution

Respondents indicated that their social security was very fragile. In enormously uneven societies such as Odisha, women with disabilities experienced multiple layers of disempowerment and prejudice, underdevelopment and patterns of structural disparity that continued and strengthened the systemic forms of poverty in rural villages. The findings revealed that the intersection of gender, disability and social exclusion reinforced the drawbacks of disabled women in terms of their exclusion from access, opportunities and resources. It is already discussed elsewhere that the general socio-economic conditions of the disabled women were particularly bleak due to lack of basic skills, restricted educational opportunities and limited livelihood/employment options. Moreover, regarding the coping mechanism of Dalit disabled women, this study indicated that they faced additional layers of exclusion and were disproportionately affected by the cyclone making them stereotypes, unproductive and chronic underdevelopment.

As reported by most of the respondents, access to relief and other post-disaster support services varied significantly among different people. There were social dynamics within a particular family that decided who is going to receive relief materials. This study revealed that women with disabilities were both overlooked and marginalized during the emergency and relief phase corroborating the findings of numerous international studies which have stated that 'persons with disabilities are too often neglected in the contingency planning, assessment, design, and delivery of humanitarian relief' (Handicap International 2015: 5). This study also pointed out that discrepancies and leaving out of various groups and individual families were less common in areas where local relief coordination mechanisms existed, and that these discrepancies were severe in areas where these mechanisms were not developed and fewer materials received from outside. Citing the responsibility of village leaders, all the respondents from Ersama village said that successful and unbiased allocation of relief materials were carried out by their community leaders who acted as coordinators by ensuring that the relief is distributed to all of them.

Violence against Women with Disabilities

All respondents, particularly the respondents with vision disability, hearing disability and lower limb disability, reported that they experienced physical, psychological, financial and sexual abuse in a domestic or intimate setting that are perpetrated by partners, family members or by other members of society. During the recovery stage, they faced mental abuse by healthcare providers or caregivers. Many of them narrated their story stating that they are almost predestined victims of male violence because of their social status and occupation. In many cases, they felt scared to lodge complaints against their abusers as they relied on them for care. Regarding physical abuse, they explained that most of them were handled roughly in every disaster evacuation process and were asked to stand for an intolerable length of time for getting relief and aid from governmental and non-governmental organizations. They also faced sexual abuse in different settings such as demanding or expecting sexual activities in return for help or being left naked or exposed. Respondents also illustrated the financial abuse in terms of not able to get money for their medicines or other basic expenditures during the emergency and recovery stages in several other disasters after the super cyclone. In rare cases, some of them noted how personal assistance providers stole money and individual articles. Some scheduled caste respondents stated that they were experiencing abuse in connection with the support needed for their disability, including medication abuse. Medication abuse consists of being given too much, too little or denied essential medication.

Several respondents with a higher educational background reported that they have been experiencing moral violence as the common people didn't show any interest in recognizing their importance or talent to develop the society. They also wanted to compare between disabled women and other women as well as between disabled women and disabled men. It is worth mentioning to note that though disabled women faced multiple vulnerabilities during normal times as well as during natural disasters, very few effective policies were designed for them. However, other women got remarkable attention from every corner of the society, every law supported women and every service to some extent improved the quality of their lives which systematically left out any reference to disabled women. As far as the comparison between disabled men and disabled women are concerned, they opined that women in their category faced multiple hazards and violence which men did not. Disabled women are the soft target of sexual abuse, domestic abuse and exploitation. The chances of physical and emotional abuses are high and as a result they face double exploitation. Disabled women suffered more due to lack of access to resources, poverty and isolation. Thus, the double discrimination denotes that the experiences of disabled women are very much different from those of disabled men. Boylan (1991: 1) presents a similar analysis concerning the disabled women stating that 'disability diminishes sharply their often inferior roles, even in their own households. The stigma of disability, with its myths and fears, increases their social isolation. When no rehabilitation facilities are available, they become immobile and housebound, and their isolation is complete'. Ranju Bala, aged 27, of *Bhoi* (Harijan) caste with cerebral palsy, narrates:

> I developed a permanent movement disorder from my childhood. I am sorry to say that my rights to marriage, sexual and reproductive rights, and family life are denied because I am a disabled woman; therefore I can no longer fulfil the roles of a sexual partner, mother and care-giver. Further, my parents advised me not to go to school due to my limited mobility and lack of financial resources. As a disabled woman, I am given low priority. If any problem occurs in my home, I am blamed for the social stigma of disability. But if you see other disabled men in my village, you will find that they are settled and happy with their married life. This is where disabled women are discriminated. Don't think that I am mentally weak to prepare myself for any type of disaster. However, my problem is rooted in physical, social and economic dependency.

Sujata Jena, aged 59, a widow with spinal cord disability, explains:

> My life has come to a standstill after the cyclone. I lost my husband after 10 years of my marriage, just before the super cyclone. Destiny has been so cruel since that time, more particularly after the cyclone. I have a daughter who is in her early teens. It is really difficult for anybody to live without a male member in our society ... The fact is that I face the bereavement and the burden of being a single parent. The patriarchal system is clearly visible in my village ... I am also considered as inauspicious after the death of my husband. The super cyclone wreaked mine and my daughter's lives leading to increased distress and vulnerability. I have become half-mad at present and am only surviving for my daughter. There is no one here to help us. The cyclone has not only imposed a huge burden on us physically but also in terms of immense psychological trauma. I don't think of any improvement in the near future.

Poverty/Lack of Access to Resources

The subject of disability in general and women with disabilities in particular is a development issue due to its direct connection with poverty (WHO 2011). There is no doubt that disability increases poverty and vice versa. Research across the world signifies that individuals with disabilities and their families are more prone to face economic and social disadvantage than those without disability (Sen 2009). Jenkins and Rigg (2003) point out that the context of disability aggravates social and economic well-being and poverty through a whole host of means including the adverse impact on education, employment, earnings and increased expenditures related to disability. The respondents with single headed households revealed that they could not send their children to schools due to limited transport and lack of access to resources. Rima Mohanty, aged 45, a widow with manual dexterity, says:

> My husband died in the super cyclone. I have three children, all of them are earning as daily labourers. Because of high costs, I could not send them to school. I could have easily sent them to school for education, had the government supported us in this regard. As a result, employment and income outcomes appear to get worse with the severity of my disability. I feel disappointed that at present I am not able to promote self-employment for my children and livelihood activities.

However, the opinion of Ahalya Bai, aged 29, a school teacher with rheumatoid arthritis, is interesting to note here:

> Those who cannot do anything for themselves or society are disabled. ... I am not disabled. I want to earn money for myself and feed my kids .... As long as I am alive I cannot accept anyone's *adesh* (suggestion). *Mu kahari bojha hebaku pusand karena* (I don't want to be a burden on others). *Jete parjanta mu school re chakiri karuthibi, batasa amara kichhi kari paribani* (Till I work in the school as a teacher, cyclones cannot harm me or our family).

Those women who were employed before the super cyclone continued to work again after the disaster. It is due to their income that problems were less visible in comparison to those who did not have any income. Similarly, some women visualized employment from home, including sewing or giving *Purana* lessons to children. These women claimed that they were feeling happy as they slowly gained control over their lives. In spite of that, the larger picture is that shortage of resources in several disaster events had strengthened discriminatory attitudes and behaviour within families and communities, leading to further exclusion.

Conclusion: The Road Ahead

The preceding result and discussion shows that women with disability experienced significant barriers regarding communication, cultural, economic, environmental, institutional, political, social and attitudinal support before the super cyclone that have further hampered and aggravated their ability to cope with daily lives as well as multiple disasters. They showed little resilience except depending on others at multiple levels with diverse degrees. To that extent, a sense of tradition and present political–economy had deeply impacted on their adaptability. Findings show that women with disability were not part of a homogenous category as multiple issues consistently weakened their needs and adaptability in keeping up with changing times. A complex interplay of caste, class, gender and disability mediated by the circumstances had emerged during multiple disasters. Thus, the departure point is that though super cyclones or other natural disasters served as catalysts, the exact sources of disasters lay in societal-induced susceptibility. Indeed, when there is segregation from enjoying the right to equality, life with dignity and respect for a woman's integrity equally with others, there is human 'disaster waiting to happen', and while women with disability live amongst the poorest of the poor, they become the worst affected. Therefore, the coping mechanisms of women with disability needs attention not only to manage hazardous events, for instances, disaster events, their immediate impacts and subsequent response and recovery activities but also to identify the manner in which pre-existing social barriers, the political economy of specific geographical areas and lack of political will to implement the 'Rights of Persons with Disabilities Act, 2016', shape the life chances of disaster victims. My contention here is that comprehending the association between women with disability and disaster broadly involves an understanding of the connection between their rights and entitlements.

Where do we move from here? A comprehensive mechanism involving all stakeholders is the need of the hour. It is the responsibility of family members, community members and society as a whole including the political willpower to provide 'high support' in the areas of physical, psychological, economical and so on in terms of allowing women with disability to take free decisions and contribute in all areas of life including education, employment, family and community life and treatment and therapy. It is high time to implement the 'Rights of Persons with Disabilities Act, 2016', with true spirit and recognize the status of disabled women with much sincerity and dedication. That is possible, only when the central and state government truly follows the principles of 'the United Nations Convention on the Rights of Persons with Disabilities', adopted by the United Nations General Assembly Convention on the Rights of Persons with Disabilities on the 13 December 2006:

(a) respect for inherent dignity, individual autonomy including the freedom to make one's own choices, and independence of persons;
(b) non-discrimination;
(c) full and effective participation and inclusion in society;
(d) respect for difference and acceptance of persons with disabilities as part of human diversity and humanity;
(e) equality of opportunity;
(f) accessibility;
(g) equality between men and women;
(h) respect for the evolving capacities of children with disabilities and respect for the right of children with disabilities to preserve their identities (The Gazette of India 2016: 1–2)

Revisiting the Odisha super cyclone of 1999 underscores the importance of further research on this topic in order to save the population most at risk during their everyday life and emergency situation. The study of social science research therefore is imperative to envisage more inclusive alternatives with a greater understanding of the social model.

References

Albert, B., Mcbride, R., and Seddon, D. (2002). *Perspectives on Disability, Poverty and Technology: A Report to Health Link Worldwide and GIC Ltd.* Norwich: University of East Anglia.

Bhatta, B. B. (1997). *The Natural Calamities of Orissa in the 19th Century.* New Delhi: Commonwealth Publisher.

Blaikie, P., Cannon, T., Davis, I., and Wisner, B. (1994). *At Risk: Natural Hazards, People's Vulnerability, and Disasters*. New York: Routledge.

Boylan, E. (1991). *Women and Disability*. London: Zed Books.

Census of India (2001). *District Census Handbook, Jagatsinghpur*. Odisha: Directorate of Census Operations Odisha.

Census of India (2011). *District Census Handbook, Jagatsinghpur*. Directorate of Census Operations Odisha.

Census of Odisha (2011). *Executive Summary*. Census of India. Available at http://www.censusindia.gov.in/2011census/PCA/PCA_Highlights/pca_highlights_file/Odisha/Executive_Summary.pdf (accessed on September 12, 2017).

Colebatch, H. K. (1998). *Policy*. Buckingham: Open University Press.

Comprehensive District Plan (CDP). (2015–16) for *Jagatsinghpur District*. Available at http://jagatsinghpur.nic.in/dpmu/CDP_jspur2015.pdf (Accessed on September 12, 2017).

Deshpande, A. (2002). "Assets Versus Autonomy? The Changing Face of the Gender-Caste Overlap in India", *Feminist Economics*, Vol. 8, No. 2, pp. 19–35.

Emmett, Tony and Alant, Erna (2006). "Women and Disability: Exploring the Interface of Multiple Disadvantage", *Development Southern Africa*, Vol. 23, No. 4, pp. 445–460. doi: 10.1080/03768350600927144

Enarson, E. and Morrow, B. (1997). "A Gendered Perspective: The Voices of Women", in *Hurricane Andrew: Ethnicity, Gender and the Sociology of Disasters*, ed. W. Peacock, B. Morrow, and H. Gladwin. New York: Routledge. pp. 116–140.

Ghai, A. (2002). "Disabled Women: An Excluded Agenda of Indian Feminism", *Hypatia*, Vol. 17, No. 3, pp. 49–66.

Government of India (GOI) (2002). *Tenth Five-year Plan (2002–2007)*. New Delhi: Planning Commission of India.

Government of Odisha (2002). *Managing Disasters in Orissa: Background, Challenges and Perspectives*. Bhubaneswar: Orissa State Disaster Mitigation Authority Publications.

Government of Orissa (1990). *Orissa State Gazetteer: Orissa State*. Vol. 1. Bhubaneswar: Government of Orissa Publication.

Gupta, D. (2004). "Introduction. The Certitudes of Caste: When Identity Trumps Hierarchy", *Contributions to Indian Sociology*, Vol. 38, Nos. ½, pp. V–XV.

Handicap International (2015). *Disability in Humanitarian Contexts: Views from Affected People and Field Organizations*. Handicap International Publication.

IFRC (International Federation of Red Cross and Red Crescent Societies) (2000). *In the Wake of Devastation: India–Orissa Cyclone*. Insight. Geneva: IFRC.

Jenkins, S. P. and Rigg, J. A. (2003). *Disability and Disadvantage: Selection, Onset and Duration Effects*. London: London School of Economics, Centre for Analysis of Social Exclusion, (CASE paper 74).

Morris, J. (1993). *Pride against Prejudice: Transforming Attitudes to Disability*. London: The Women's Press.

Nagata, K. K. (2003). "Gender and Disability in the Arab Region: The Challenges in the New Millennium", *Asia Pacific Disability Rehabilitation Journal*, Vol. 14, pp. 10–17.

OSDMA (2001). *Lessons Learnt: Orissa Floods 2001*. Proceedings of the workshop organised by Orissa State Disaster Mitigation Authority, Gopabandhu Academy of Administration, Bhubaneswar, in collaboration with Gopabandhu Academy of Administration and UN Development Programme.

Oxfam (2000). *India Disasters Report*. New Delhi: Oxford University Press.

Parida, P. K (2002). "Towards Rebuilding a Post-Disaster Society: A Case Study of Super-Cyclone Affected Coastal Orissa", *The Indian Journal of Social Work*, Vol. 63, No. 2, pp. 243–262.

Parida. P. K. (2008). "Super Cyclone Affected Coastal Orissa: A Social Vulnerability Approach", *Review of Development and Change*, Vol. XIII, No. 2, pp. 159–180.

Patnaik, N. (1969). *Caste and Social Change*. Hyderabad: National Institute of Community Development.

Planning Commission of India (2004). *State Human Development Report*, Orissa.

Pradhan, R. P. (2003). "Agriculture Efficiency of Orissa: Problems and Prospects", *Journal of Rural Development*, Vol. 22, No. 2, pp. 220–247.

Rao, I. (2004). *Equity to Women with Disabilities in India*. Bangalore: CBR Network.

Sen, A. (2009). *The Idea of Justice*. Cambridge: Belknap Press of Harvard University Press.

Shklovski, I., Burke, M., Kiesler, S., and Kraut, R. (2010). "Technology Adoption and Use in the Aftermath of Hurricane Katrina in New Orleans", *American Behavioural Scientist*, Vol. 53, No. (8), pp. 1228–1246.

The Gazette of India (2016). *The Gazette of India,* Ministry of Law and Justice. New Delhi: Legislative Department, December, p. 28. Available at http://www.disabilityaffairs.gov.in/upload/uploadfiles/files/RPWD%20ACT%202016.pdf (Accessed on September 15, 2017).

Traustadottir, R. (1990). *Obstacles to Equality: The Double Discrimination of Women with Disabilities*. Dawn Ontario: Disabled Women's Network.

United Nations (2006). "United Nations Convention on the Rights of Persons with Disabilities, Article 11", Division of Social Policy and Development Disability. Available at http://www.un.org/disabilities/documents/convention/convention_accessible_pdf.pdf (Accessed on September 22, 2017). See also http://www.un.org/esa/socdev/enable/rights/convtexte.htm (December 6, 2006).

Wasserman, David, Asch, Adrienne, Blustein, Jeffrey, and Putnam, Daniel (2016). *Disability: Definitions, Models, Experience, Stanford Encyclopaedia of Philosophy*. Available at https://plato.stanford.edu/entries/disability/ (Accessed on September 11, 2017).

WHO (World Health Organization) (2001). *International Classification of Functioning, Disability and Health* (ICF), Geneva: World Health Organization.

WHO (2011). *World Report on Disability 2011*. Geneva: WHO Press.

Wisner, B., Blaikie, P., Cannon, T., and Davis, I. (2004). *At Risk: Natural Hazards, People's Vulnerability, and Disaster* (2nd ed.). London: Routledge.

7

Impact of Nepal Earthquake of 2015 on Women

Challenges and Responses

Rishi Gupta

Assessing the Impact

Nepal suffered two major earthquakes on 25 April and 18 May 2015, of the magnitudes 7.6 and 6.8 respectively (National Seismological Centre 2015). As of 26 November 2015, It left 8,962 dead and 22,302 injured (as of 26 November) (Nepal Disaster Risk Reduction Portal 2016). Due to its epicentre in the Gorkha District, it was called Gorkha Earthquake. The earthquakes were followed by 474 numbers of aftershocks with a local magnitude of 4 (as of 17 October 2016) (National Seismological Centre 2015). Nepal witnessed earthquake of this high magnitude after 80 years (National Planning Commission 2015). The last earthquake of such a high level occurred in 1934 which had resulted in more than 10,000 deaths in the Kathmandu Valley (National Planning Commission 2015). The 2015 earthquake destroyed several historical sites including UNESCO recognized *Dharahara* (also called *Bhimsen* Tower) and damaged Buddhist stupas of *Swayambhunath* and *Boudhanath,* the *Durbar* Squares of Kathmandu, Patan and Bhaktapur (BBC 2015).

Out of a total of 75 districts, 14 districts were found most 'crisis hit' (National Planning Commission 2015; National Seismological Centre 2015). The massive damage caused by the earthquake includes 775,782 fully damaged houses and 302,774 partially damaged houses (as of 26 November 2016) (Nepal Disaster Risk Reduction Portal 2016). According to Post-Disaster Needs Assessment (PDNA) 2015, the total value of disaster effects (damages and losses) is NPR 706 billion/US $7 billion which is more than 100 per cent of the gross fixed capital formation for FY 2013–2014 (National Planning Commission 2015).

An expedient assessment of the 2015 earthquake shows that 14 most affected districts[1] include about 2,710,239 women (50.5% of population), 327,000 female-headed households (26.5% of households), 40,000 women with disabilities, 163,000 women over the age of 65 years (6% of female population), 764,000 girls aged 14 or under (28% of female population), and 795,000 women and girls who are illiterate (29% of female population) (United Nations Office for the Coordination of Humanitarian Affairs 2016). A review of studies conducted in the past on vulnerability of women in the event of a natural disaster shows an incessant ratio of women are the most affected than their male counterparts. Especially, poor women since they have lower preparedness, resulting from the lack of information and resources (Kusakabe and Veena 2015; Pincha 2008). Additionally, the structural damages caused by the natural disasters and the chaotic social breakdown make women uniquely vulnerable to sexual abuse and domestic violence. Ignorance of such violence against women (VAW) by the authorities further increases the amount of violence (Bimal 2011; Ramdas and Chew 2005; Solotaroff and Rohini 2014; Stoltman et al. 2007). PDNA categorically mentions that the effect of the earthquake was high on women 'partly because of gendered roles that disproportionately assign indoor chores to women' (National Planning Commission 2015).

The earthquake highlighted the aspects of inequalities in the society spanning geography, income and gender (National Planning Commission 2015). The decade-long old civil war in Nepal from 1996 to 2006, which led to the collapse of the centuries old feudal monarchical rule in Nepal had seen an active participation of the women (Asian Development Bank 2010). The Maoist waged war against the monarchy was able to attract women due to their ideological support to gender and class equality (Einsiedel et al. 2012) but the vulnerability of women during earthquake wide opened the outcomes of the civil war in this regard.

In handling the post-Earthquake situation in Nepal, a pale response could be seen on the part of the administration. As Ramdas and Chew (2005) of Global Fund for Women argue that availability of health needs bring women on the most disadvantaged side. 'Pregnant women lack obstetrical care and may miscarry or deliver under extremely unsanitary conditions. Displaced women often lack access to contraception and even sanitary supplies, which in some cases have been doled out by men' (Ramdas and Chew 2005). Also, displacement of women and absence of male due to the high ratio of migration for livelihood causes severe pain since displacement leads women to temporary shelters and open spaces. This adds concerns over the chances of VAW as was reported in Nepal (National Disaster Risk Reduction Centre 2016; United Nations in Nepal 2016). Nepal earthquake saw an extra workload on women in the districts such as Gorkha, one of the most affected districts, which has a skewed female population due to male outmigration, which means women will take on a larger responsibility of rebuilding sectors such as agriculture and livestock (National Planning Commission 2015).

Issues and Response

Earthquakes are the most destructive natural disasters and can produce several types of loss, including physical, socio-economic and cultural. While the loss of lives leaves an imprint for future cautions and readiness, damage to the infrastructure costs the future hardships in the form of health, home and livelihood. An earthquake takes the country years backwards from the developments made in the form of the infrastructure and economy (Shaw and Kobayashi 2001). The economic well-being of the country is one of the very crucial elements in deciding the duration that an affected country may take to recover. On similar lines, the 2015 earthquake in Nepal might take a decade or so to bring life to the normal.

The prime motive of the present study is to assess the impact of the 2015 earthquake on women. This study is the further addition to the knowledge of the event. This study assesses the timeline of issues that emerged in the post-earthquake phase and their impact on women. Major issues include health status of women, scarcity of medical supplies, financial needs and discriminatory gender-based political representation. This study uses facts and figures provided by the government, non-governmental organizations (NGOs) and international non-governmental organizations (I/NGOs) to set up the case. Several personal interactions with the researchers working on Nepal has benefitted in collecting information on the subject. Finally, this study is a result of my personal follow-up of the events in Nepal. Assessment of the 2015 earthquake and its implications on women are followed by recommendations and conclusion.

Health

'Disasters are windows that showcase the prevailing gender-based inequalities in the society' (Ariyabandu 2009: 11). The aftermath of 2015 earthquake in Nepal is no less than a reflection of prevailing gender-based discrimination and inequality. Discrimination in health facilities, education, political representation, labour market and different treatment towards right to citizenship as provided to men are few of the serious challenges faced by women in Nepal today (UNDP 2016).

Nepal ranked 145 in the United Nations Human Development (UNHD) report of 2015, just below Kenya. It ranked 108 in the Gender Inequality Index (GII), becoming second on top in Low Human Development (LHD) countries. In maternal mortality ratio (MMR), Nepal has 190 deaths per 100,000 live births and 73.7 per cent per 1,000 adolescent aged between 15 and 18 (UNDP 2015). With a GDP of US $20.88 billion (FY 2015) (Trading Economics 2016), Nepal for past several decades has been depending on the foreign aid and line of credit for its annual budget expenditure (Ministry of Finance 2015; Vanitha and Ashokan 2016). The annual expenditure on health for the financial year (FY) 2014–2015 accounted 5.8 per cent of the total budget expenditure, which included total expenditure on health of US $137 per capita (World Trade Organization 2016). The earthquake affected Nepal's annual budget spending on health. Initial figures estimated that an additional amount of US $147 million for health facilities was required with an additional amount of US $50 million for nutrition (National Planning Commission 2015).

The tremor required immediate health relief. The damage caused to health facilities included 4,904 fully and 1,159 partially damaged. It made the situation critical, especially in the rural and hilly areas. Healthcare units were operated under the temporary shelters. Women had to defecate in open spaces and use unfiltered water, which increased the chances of physical harm and VAW. Also, due to the overwhelming need for food, shelter and water, menstrual health had fallen on the list of priorities during the initial disaster relief management activities (*New York Times* 2015).

PDNA estimated that 72,784 pregnant and 61,715 lactating women were affected by the earthquake in the 14 most affected districts. The report further estimated that:

> 10,327 babies are born every month in the 14 most affected districts and due to loss of the health infrastructure, there might be a surge in the complications during pregnancies and required basic health care might not reach to them. (National Planning Commission 2015)

Tomoo Hozumi, The United Nations Children's Fund (UNICEF) representative of Nepal, added that around 90 women were in need of emergency surgeries in affected parts of Nepal in a day and 'presently, the systems cannot cope' (UNICEF 2015). Commenting on the health infrastructure in Nepal, he further added:

> Despite the progress made over the last few decades, access to quality maternal and newborn healthcare in Nepal was still limited even before the earthquakes. Every day, 38 newborns died largely of preventable causes, and one woman died every eight hours from complications during delivery. Prior to the quakes, just over a third (38.64 per cent) of mothers in the 14 worst-affected districts gave birth in a recognised health facility, while around three in five pregnant women across the country (59.5 per cent) received antenatal services. (UNICEF 2015)

Following the earthquake and its aftershocks, women and newborn babies were a critical aspect of the immediate help and rescue. Lack of resources and infrastructure on government's part remained one of the worries for the organizations working with the Nepal government to provide healthcare services on the ground in the affected areas. In the South Asian subcontinent, mental health is often replaced with words such as 'evil', 'cruel', 'impulsive' and 'quarrel-loving'. Nepal is no different in using these words. Most of the rural population believes that mental illness is the result of bad fortune and people visit local faith healers for a cure. There is no concept of biological causes of mental illness, even among the educated (Regmi et al. 2004). In an alarming analysis, World Health Organization (WHO 2015) notes that:

> 450 million people suffer from mental, neurological, behavioural or substance use disorders worldwide and that one in every four persons will be affected by a mental and neurological disorder at in some stage in life. Four out of five people in low- and middle-income countries that need services for mental and neurological and substance use disorders do not receive it.

Examples of neglect towards mental and psychological health are seen in several cases of natural disasters in the past around the world. The 2001 Gujarat earthquake in India, 2004, tsunami in India (WHO 2005) and 2010 flood in Pakistan (Naiz 2010) had massively affected women's mental health. To this day, several parts of these countries lack proper counselling centres, if any, the rural areas have none. The 2004 tsunami had hit several coastal regions of India. The tsunami had primarily paralysed women, closely involved in the fishing business. Houses were destroyed, and several had lost their families, and several cases went unreported. WHO's initial public health risk assessment report on 2015 earthquake in Nepal stated that the present existing mechanism of community-based mental health programmes and free medications, run by the governmental and NGOs, had the capacity to address the issues in selected areas only, and the similar observations were made by the International Organization for Migration (IOM) (IOM 2017; WHO 2015). Some studies on Nepal earthquake suggest that the mental healthcare in the post-earthquake phase will have long-term emotional consequences. Feeling of powerlessness and lack of control over forces bigger than oneself will add more to the vulnerability (Bennett 2015).

WHO's Mental Health Atlas 2014 shows some alarming facts on mental healthcare in Nepal. The ratio of the psychiatrists against the total population of the country stands poorly. For every 100,000 population only 2 per cent of the total psychiatrists and a total staff of 78 personnel were reported in the country (WHO 2014). An estimated 440 in-patient beds for people with mental illness (combining both governmental, 112, and private hospital facilities, 327), which amounts to 1.5 beds per 100,000 population across the country (Luitel et al. 2015). With only one complete mental healthcare hospital in the country and that too in Kathmandu, Nepal heavily relies on the non-governmental actors in this regard (Simkhada et al. 2015). The accounted data was gathered a year before the earthquake-hit Nepal and in light of the earthquake mental health services were hardly available. Despite the availability of the policy and planning in this regard, 'mental health receives insignificant attention' (Simkhada et al. 2015: 1). Hence, more attention towards mental healthcare can help in the post-earthquake phase to address the mental healthcare in the country. Also, Nepal needs to recruit more staff and arrange more mobile services that can reach to rural and remote areas. A necessarily gendered approach needs to be taken into account while dealing with mental healthcare in the country.

The customary practices such as *Chaupadi*,[2] despite being outlawed by the Government of Nepal (GoN), have been prevalent in several parts of Nepal (*The Economist* 2014). Under this practice, menstruating women are considered impure. They are forced to live in temporary arrangements outside the house. Also, an open discussion on women's hygiene is considered improper in several societies across Nepal. In the post-earthquake phase, this was one of the alarming aspects of the women's health. Marred by poverty, several women could not afford to buy sanitary pads. Therein they were forced to use unhygienic cotton clothes which not only affected their reproductive system but also raised the chances of several infectious diseases (*New York Times* 2015). With the help of NGOs and I/NGOs, the required utility products were provided, but in several remote areas, women had to reuse old cotton pads by washing them again. Although no major incidents of such practices were reported from the makeshift tent arrangements, the unhygienic nature of living under them created further health risks to women.

While the earthquake damaged major health facilities in the country, response teams from various countries reached to the affected areas and provided health services. Various I/NGOs and countries in association with various ministries arranged and established response teams to provide assistance in the affected areas. India was one of the first countries to extend its help on the ground in Nepal by launching Operation Maitri (Friendship). A medical contingent of civilian medical doctors along with two tonnes of medical supplies (Ministry of External Affairs 2015) flew from New Delhi to Kathmandu. Medical relief was followed by 60 more countries sending their rescue and medical teams to Nepal.

This year in a bulletin, National Reconstruction Authority (NRA) reported that on priority basis out of 1,080 damaged medical facilities, 444 are under construction with the help of I/NGOs and GoN plans to complete other facilities at the earliest (NRA 2017). The report further added:

> So far about 200 health centres have begun operations using prefabricated structures and 100 have been handed over to the concerned agencies. These will be used to provide health services until the permanent structures are ready. At least 133 I/NGOs have sought approval for supporting the construction of houses, schools, health post and heritage sites in the affected districts. (NRA 2017)

In the absence of the health infrastructure, I/NGOs namely Global Fund for Women, UNICEF, CARE, European Commission, Doctors without Borders, Humanity First, Mercy Corps, Oxfam, Save the Children International Medical Corps, Red Cross Society and other local NGOs are working to provide health services in the affected areas. Initial response by UNICEF towards women healthcare included setting up of emergency and maternity tents, shelter homes for mothers of the newborn, providing iron and folic acid supplement for pregnant and new mothers in affected areas. Towards financial help, an amount of CHF 39,300,894 (Swiss Franc) was collected by the International Red Cross Society from April 2015 to December 2017 to help Nepal (IFRC 2017). India and China along with other international donors provided a grant of US $500 million and the US $50 million to rebuild health infrastructure in Nepal (NRA 2017). Japan International Cooperation Agency has contributed US $36,457 specifically for Paropakar Maternity and Women's Hospital and Reconstruction of Bir Hospital (NRA 2017). While reconstruction work of the healthcare units is in progress, GoN must ensure that the new constructions follow infrastructural expertise to minimize future hardships.

Landownership

Nepal has a destitute track record of landownership among women. Discrimination by gender is deep-rooted and widespread in Nepal irrespective of caste and age. In parts of Nepal, patriarchal traditions remain strong along with the widespread violence and discrimination against women. According to Community Self Reliance Centre (CSRC) (2009), women run the household and do much of the fieldwork, but have a low status in society and no right to property (CSRC 2009). With least in their hands, millions of women are forced to work as labourers in other fields on minimum wages. Landlessness affords no status in the community and disenfranchises millions from their basic human rights (CSRC 2009). Only 19.7 per cent of women in Nepal own around 5 per cent of the land throughout Nepal, and around 11 per cent have effective control over their property (IOM 2016). GoN has introduced schemes of tax exemption on a land purchased and registered under the female members of the family or in many cases exemption of 25 to 30 per cent tax if registered jointly in the name of husband and wife but this has not produced desired results. The pattern of owning land shows that women are much less likely than men to inherit land, to have land registered in their name or to obtain documentation to prove their entitlement (Oxfam 2016).

Political will to empower women based on landownership, including the control over land and property, can help them in the reconstruction of houses. A report by IOM on *Gender and Migration* states that:

> Women constitute 51.5 percent of Nepal's total population, and around 75 percent of women are engaged in agriculture as their primary occupation, it is ironic to find that women often don't have ownership of land that they have been tilling for years. (International Organization for Migration 2016)

Before the earthquake, more than 25 per cent of Nepal's population was landless or land poor, the majority were Dalits, *Janajatis* and women (CSRC 2009) and this might go years low in post-earthquake phase. The lack of ownership of land has landed women to face problems in receiving the relief amount by the government since the recipient was subject to the owner of the land or head of the household (Israeli 2015). Hence, efforts by the GoN need to address this issue so the transfer of the relief amount can be done with ease. GoN should also run awareness programmes on registration of land in the name of female members of the family which can help in empowering women.

Political Representation

In the case of political representation, Nepal's case has improved in recent years. At the time, monarchy was in its last phase, parliament had a meagre women representation of around 6 per cent. After the first democratic elections in 2008, women representation made remarkable progress to 44 per cent in the constituent assembly (CA), but during second CA elections in 2013, it came down to 33 per cent

World Economic Forum (WEF) (2011, 2013). Presently, 30 per cent lawmakers in the CA are women, and only 14 per cent hold ministerial portfolios (WEF 2016). No woman in Nepal has ever held the office of the Prime Minister. It is for the first time that women are holding the offices of the President, Chief Justice and Speaker of the CA (Pradhan 2015; The Carter Center 2013; *The Indian Express* 2016; UNDP 2015). Despite these efforts, the existing capacity gap means that women are less likely than men to have better education, contacts and resources needed to become effective leaders (UN Women 2016). Advocating the need to make women more politically empowered, United Nations has critically analysed several challenges faced by women in claiming their political representation, which is necessary for their structural empowerment. UN notes that discriminatory laws and institutions still limit women's options to run for office.

Eight months after the 2015 earthquake, NRA was formed with an act of the parliament to 'promptly' complete the reconstruction works and 'provide social justice by making resettlement and translocation of the persons and families displaced by the earthquake' (Ministry of Law, Justice & Parliamentary Affairs 2016). Being one of the apex bodies to address the post-earthquake reconstruction and rehabilitation, it has seen a poor representation of women members since its inception. The women rights activists have been vocal in raising this issue, so women are not left out in the reconstruction process. Women's representation in NRA has also concerned the donor countries as well as international organizations, which are actively involved in the reconstruction process on the ground. Australian Embassy, UN Women Nepal and *Sancharika Samuha* Nepal in a letter to the NRA recommended the latter to provide a formal mechanism where women have a better say within the national and local level NRA structure and recommended to stop political appointments that are submissive (Australian Embassy in Nepal, UN Women Nepal and Sancharika Samuha Nepal 2016; *The Kathmandu Post* 2015c)

Whereas the poor representation of women members in the NRA questions political will of the government and political parties, the formation of NRA itself had witnessed political theatrics. Political parties had a major role in causing a delay in the formation of NRA as they could not agree on the name for the chief executive officer.

Eventually, adding further suffering on people rather than relief. On 4 August 2016, Communist Party of Nepal's (Maoist-Centre) chief Pushpa Kamal Dahal, also known as Prachanda, assumed office as the Prime Minister of Nepal. It was for the third time in a row since 2013 CA elections that a new head of the government had taken over charge as the Prime Minister of Nepal and eighth Prime Minister since the collapse of the monarchy. The previous government led by the Prime Minister K.P. Oli had appointed Sushil Gyawali as CEO of the NRA, but amid of the public and international criticism over NRA's slow pace, he was sacked from his post by Prachanda (*The Kathmandu Post* 2017a). The incompetency of NRA's ex-CEO or political vendetta might have led to his fall but this was an apt opportunity for PM Prachanda to increase the number of women members in NRA as the larger debate on women representation has remained questionable.

External Political Conundrum

Nepal is a land-locked country between its two Asian giant, India and China. Conditioned by its geographical location, its politics is highly influenced and often moulded by the two. With no access to sea, Nepal's major trade (almost 98%) with the world takes place through the nearest Indian seaports, Haldia in Kolkata and Vizag port in Andhra Pradesh, in compliance with the bilateral treaties and international laws for land-locked countries (Haidar and Bhattacherjee 2016; Taneja et al. 2015). In an understanding between Nepal and China signed in March 2016, Nepal will be able to access China's Tianjin port after the infrastructural requirements are completed but poor infrastructure and difficult terrain, this arrangement is unlikely to take place in near future. This agreement with China was signed after the Madhesi people (from the southern plains in Nepal) had held a month-long protest against the GoN for not addressing their problems in newly implemented constitution in November 2015, which eventually blocked the major trade transit points between India and Nepal.

At the time, Nepal required an uninterrupted supply of goods, especially medical and petroleum products, chaos and blame game had begun to dominate the political space, which not only neglected the needs of the hour instead, the focus shifted towards India for its alleged role in supporting the month-long blockade. However, India staunchly denied having any role in supporting the protesters and called it as an internal matter of the country. This blame game drastically affected the relations between Nepal and India. The state's failure in providing welfare to people has often been seen through the prism of external players, mostly India and China. But, present day Nepal is no different than it rested during the earthquake.

Presently, Nepal shares a very complex relationship with India. In this light, to avoid any future debacles in the hour of need, Nepal needs to come up with some alternative policy to avoid external pressure and infrastructure-related arrangement. Petrol, medicines and cooking gas are few of the primary needs and their halt, in any case, have an impact on the people of Nepal. Under the temporary shelters, women are forced to cook with the wood which eventually affects their health. Hence, the GoN shall store sufficient supplies of emergency products and a timely review of the availability must be a priority.

Migration

International migration in the last few decades has become one of the most sought after means of livelihood in Nepal. People are moving away from working in the crop fields and willing to work abroad (Kaphle 2015). There is a significant rise in labour migration from Nepal to India, Malaysia, Gulf and South East Asian countries, contributing large sums to Nepal's GDP in the form of remittances. In the 2014–2015 (FY), the inflow of remittances stood NPR 589.6 billion accounting 27.7 per cent of Nepal's GDP compare to NPR 58.6 billion in the 2003–2004 (FY), which accounted 10.9 per cent of Nepal's GDP (Ministry of Labour and Employment 2016). These figures might be much higher since migration to India is mostly unorganized and undocumented. There is no mechanism to keep records of the people moving from Nepal to India due to an open border and no visa policy between the two countries (Limbu and Sijapati 2012). In the last two decades, 14 per cent population/3.8 million of the current population has migrated out of Nepal, citing employment as the reason (Ministry of Labour and Employment 2016).

In the 2014–2015 (FY), out of a total of 34,384 international migrants, male migration accounted 31,198 in number excluding migration to India (Ministry of Labour and Employment 2016). Some of the districts in Nepal have seen a rising pattern of international migration, making women take charge of the family, farm and post-earthquake reconstruction work. Meagre representation of women in decision-making matters in the family, political sphere and administrative domains is one of the common phenomena in Nepal (Valdés 2009). Post-earthquake, absence of male population immensely affected the reconstruction work. The impact has been high 'possibly because of the "hard labour" entailed during the rescue, relief and evacuation processes, following any disaster' (Sijapati et el. 2015: 17) There is a uniformity of outcomes in the literature showing negative impact of natural disasters on women due to male migration.

The GoN had issued directions to several of its diplomatic missions abroad, especially to Malaysia and Gulf countries, to request the host governments to allow Nepali migrant workers to return home in the wake of the devastation caused by the Great Earthquake which could help the families in reconstruction works (*The Kathmandu Post* 2015b). Several districts, that is, Gorkha and Lamjung, have a very high rate of the male labour migration, but in many cases, host countries did not respond well to the request (*The Kathmandu Post* 2015a). Issues such as *Kafala* sponsorship system by the Gulf countries, which allows an employer to retain passports and other necessary documents of the hired migrant labour, minimized their chances of return. Many migrants preferred to stay back because they paid hefty amounts in getting jobs to the employment agencies, and returning would not have helped them in paying loans that they took before their departure from Nepal. Although returning from Gulf countries was costly, many of migrant workers from India returned home after the quake. Also, migration within the country saw a similar pattern of their return. Workers from other districts left the quake-affected districts in fear, while a reverse trend of workers leaving their workplaces in non-affected areas to help their families in quake-affected areas also took place (National Planning Commission 2015).

Studies have found that the impact of male outmigration on the family members, especially on women's workload, depends on the ability to hire labour from remittances earned, and the shortage of labour failed women miserably. Migrant households have also been facing the problem of assessing the relief support provided by the GoN and other NGOs (Maharjan et al. 2012). To access government support, the lack of legal documents proved a major obstacle. People who had lost their houses also lost their citizenship and other documents along with them. It was very difficult for them to retrieve or replace these and access government relief schemes. For non-governmental support, migrant households were more vulnerable than non-migrant households due to a lack of able-bodied members to collect support materials (Maharjan et al. 2016). When disasters happen, such as the 2015 earthquake in Nepal, lack of men in the village can put communities in further danger (Jena 2017). Male migration from Nepal has left women with an exhausting burden on their shoulders.

Financial Support

On 25 June 2015, GoN organized the International Conference on Nepal's Reconstruction (ICNR) and used the platform to disseminate PDNA findings to seek technical and financial support for reconstruction. India, China and Asian Development Bank topped in pledging financial support to Nepal, followed by other countries and I/NGOs. Although the pledged amount was USD 4.1 billion, the adjusted (actual) was the only USD 3.43 billion as the USD 670 million was for regular programmes. Of USD 3.43 billion pledged for reconstruction and recovery, USD 2.7 billion (79.3% of pledge) has been confirmed through MOUs. United States (USD 9.6 million), Swiss Development Cooperation (USD 9.0 million), Canada (Canadian Dollars 10 million) and other donors have also contributed to Multi-Donor Trust Fund (NRA 2017).

Criticized for its 'snail pace' reconstruction process, GoN has begun distributing grants for house construction damaged due to the earthquake in the 14 most affected districts. Grant distribution is done under cluster system based prioritization of the most affected and less affected districts. The housing grant in the 14 earthquake-affected districts showed that 77.13 per cent of the total 626,036 eligible beneficiaries received the first instalment (as of 19 January 2017). As per Central Level Project Implementation Unit (CLPIU) under the Ministry of Federal Affairs and Local Development (MoFALD), 86.45 per cent (459,887) beneficiary households under Cluster 1 in the 11 worst-hit districts outside Kathmandu Valley received the first instalment of the grant. Distribution of the grant in the three districts within the valley categorized under Cluster 2 had also received the first instalment (NRA 2017; *The Kathmandu Post* 2017b).While the pace of the relief amount remained slow, single women and women with disability faced problems in receiving relief amount. Financial relief on priority basis could help their livelihood since the majority of them do not own a land for agricultural purpose. Hence, a mechanism should be established, so even after two years of the tragic event, household women with disability are not left behind in receiving financial support from the government.

Discussion and Conclusion

The 2015 Nepal earthquake brought the normal public life to a standstill. An earthquake of 7.6 magnitude followed by several aftershocks caused unwelcome destruction. While natural disasters see no gender but in the case of Nepal women were found more affected and helpless. A surge rise in VAW, trafficking, sexual abuse, gynaecological damages and reproductive damages are few of the challenges that emerged in post-earthquake phase. While deep-rooted gender discrimination made it inevitably impossible for single mothers, widows and displaced women to receive prompt relief, existing societal structure further exposed the invariable social treatment received by the women on a daily basis. While the unfortunate natural occurrence questioned the state readiness and response for a natural disaster, it further exhibited the political will of the lawmakers and their collective failure in addressing the problem of reconstruction and relief. Till the date, affected population is the victim of the political deadlocks, which has eventually delayed the relief process. The economic strength of the country has shown the least to offer financial stability to women which can facilitate steady reconstruction process.

Taking literature into account on earthquake and its gendered impact, there are no doubts that women suffer more than their male counterparts. The present study found that in post-earthquake phase, women's condition has worsened due to the existing gender gap, poor health infrastructure, political mainstreaming and above all neglect towards the planning and preparedness to address any natural disaster. To conclude, firstly, the geographical and critical seismic location of Nepal should be taken as the matter of national security. It is to note that

> Nepal is one of the 20 most disaster-prone countries in the world. More people are killed by disasters in Nepal than in any other country in South Asia, making Nepal extremely vulnerable to different types of natural disasters like earthquakes, floods, landslides, fire, drought and others. (Shakya 2016: 75).

Hence this area requires special attention. In the year of 2008, GoN came up with National Strategy for Disaster Risk Management (NSDRM). The strategy was to provide planned guidelines in a natural disaster situation (Government of Nepal 2008). Despite NSDRM providing a clear framework, 2015 earthquake saw directionless activities in providing relief (Adhikari et al. 2016). Although in policy formulation a gendered approach is taken into account, it is equally ignored at the time of implementation. While the state preparedness in facing a natural disaster was non-existing, outcomes of the earthquake coupled with the present political situation in Nepal raise the warning alarms for future.

Secondly, the public health infrastructure in Nepal demands a serious review. Sufficient numbers of the skilled health staff should be made available so that the critical aspects of the post-earthquake reproductive complications among women can be addressed at the earliest. Further, the 2015 earthquake should be taken into account while formulating the future public health policies. The government should also increase the number of awareness programmes to address the discriminatory practices such as *Chaupadi* across the country. General hygiene must be given priority in hospitals, and the same should be conveyed to the young generation. Access to medicines should be affordable so people irrespective of their financial condition can access them with ease. During this study, services to mental healthcare in Nepal was found insufficient in the post-earthquake phase. GoN needs to work in collaboration with I/NGOs so that the prompt relief towards mental healthcare can be made available in the country. Preventative measures need to be strengthened to address VAW in normal and emergency situations.

Thirdly, the low rate of landownership among women is yet another hurdle in their financial empowerment. Awareness programmes to provide legal advice to women should be a priority for the law and governance of the country. Despite making substantial progress in the field of political mainstreaming of the women, their voices have either been ignored or do not receive proper attention. Ministries such as Ministry of Women, Children and Social Welfare (MWCSW) require a gendered approach while appointing a minister. Also, it is unfortunate that the NRA did not pay heed to the domestic and international criticism for not taking women members on board. Incidents like this can be avoided so that a positive input by the women members can be made available on vulnerabilities faced by them during and post-earthquake.

Having fought a decade-long war against the feudal forces, Nepal has emerged as one of the newest democracies in the twenty-first century. Conduct of regular CA elections and promulgation of a written constitution are few of the recent achievements made by Nepal since 2008 despite several political deadlocks. It is commendable for the country to have achieved democratic values by providing fundamental rights to its people but at the sometime it has failed to fill existing social, ethnic and regional divide in the country. The existing discriminatory laws concerning citizenship rights to women and huge dissatisfaction of the people from southern plains (Terai) will further expand the existing divide. Post-earthquake reconstruction phase has been crucial with regard to providing quick help to the victims as the government focus is often diverted more on the political deadlocks rather addressing the critical issues.

Notes

1. Fourteen most affected districts are Bhaktapur, Dhading, Dolakha, Gorkha, Kavrepalanchok, Kathmandu, Lalitpur, Nuwakot, Ramechhap, Rasuwa, Sindhupalchowk, Makawanpur, Sindhuli and Okhaldhunga.
2. The custom called *Chaupadi* is derived from two Nepali words: *Chau*, meaning menstruation, and *Padi*, meaning women. For a week of every month, women are considered impure when menstruating and during this period they become untouchable. They must eat their meals separately, make no physical contact with other people or with water sources and must sleep with little or no bedding. In some areas, women are expected to sleep outside in a tiny hut called a *Goth* until they are clean again. The Nepal Supreme Court banned *Chaupadi* in 2005, but government regulation hasn't reach remote western regions (Robinson 2015).

References

Adhikari, B., Mishra, S. R., and Raut, S. (2016). "Rebuilding Earthquake Struck Nepal through Community Engagement", *Frontiers in Public Health*, Vol. 4, No. 121, pp. 1–4.

Ariyabandu, Madhavi Malalgoda. (2009). "Sex, Gender and Gender Relations in Disasters", in *Women, Gender and Disaster: Global Issues and Initiatives*, ed. P. G. Dhar Chakrabarti and Elaine Enarson. New Delhi: Sage Publications.

Asian Development Bank (2010). *Overview of Gender Equality and Social Inclusion in Nepal.* Philippines: Asian Development Bank.

Australian Embassy in Nepal, UN Women Nepal and Sancharika Samuha Nepal. (2016). "2016 Report to NRA on Building Back Better and Safer for Gender Equality". Kathmandu: Australian Embassy in Nepal, 1–8. Available at http://nepal.embassy.gov.au/files/kmdu/2016%20Report%20to%20NRA%20on%20Building%20Back%20Better%20and%20Safer%20for%20Gender%20Equality.pdf (Accessed on November 13, 2016).

BBC (2015). "Nepal Earthquake: Damaged Historic Sites Reopened". 15 June. Available at http://www.bbc.com/news/world-asia-33130231 (Accessed on December 31, 2016).

Bennett, Claire (2015). "The Nepal Earthquake Have Unleased a Mental Health Disorder", *The Guardian*, 15 May. Available at https://www.theguardian.com/commentisfree/2015/may/15/nepal-earthquake-mental-health-disaster (Accessed on January 18, 2017).

Bimal, Paul Kanti (2011). *Environmental Hazards and Disasters Contexts, Perspectives and Management.* Chichester: John Wiley & Sons.

CSRC (Community Self Reliance Centre). (2009). *Land and Land Tenure Security in Nepal.* Kathmandu: Community Self-Reliance Centre.

Einsiedel, von Sebastian, Malone, M. David, and Pradhan, Suman (2012). *Nepal in Transition: From People's War to Fragile Peace.* New York: Cambridge University Press.

Government of Nepal (2008). "National Strategy for Disaster Risk Management in Nepal". Kathmandu: National Society for Earthquake Technology-Nepal, 18–19. Available at http://www.nrcs.org/sites/default/files/pro-doc/NSDRM%20Nepal.pdf (Accessed on December 30, 2017).

Haidar, Suhasini and Bhattacherjee, Kallol (2016). "Nepal Seals Agreement on Transit Rights through China" *The Hindu*, 16 March.

IFRC (International Federation for Red Cross and Red Crescent Society). (2017). "Emergency Appeal Operation Update. Earthquake Recovery Update". Kathmandu: Nepal Red Cross Society. Available at http://www.nrcs.org/updates/nepal-red-cross-earthquake-recovery-operation-updates (Accessed on January 25, 2017).

IOM (International Organization for Migration). (2016). "Securing Women's Land and Property Rights in Nepal". Available at http://www.iom.int/news/securing-womens-land-and-property-rights-nepal (Accessed on December 12, 2016).

IOM (International Organization for Migration). (2017). "IOM Nepal Earthquake Response: Health". Available at http://nepal.iom.int/jupgrade/index.php/en/earthquake-response/17-main/149-health (Accessed on January 17, 2017).

Israeli, Ella (2015). "Post-Earthquake Relief Efforts: A Case Study", *Independent Study Project (ISP)*, Vol. 19.

Jena, Manipada (2017). "Women Bearing the Strain of Climate-Forced Migration". *The Wire*. 16 January. Available at https://thewire.in/99737/women-strain-climate-migration/ (Accessed on January 17, 2017).

Kaphle, Anup (2015). "Nepal, Once Known for Farming, Now Exports People; Migrants Earn Big but Face Risks". *The Washington Post*, 24 March. Available at https://www.washingtonpost.com/world/asia_pacific/nepal-once-known-for-farming-now-exports-people-migrants-earn-big-but-face-risks/2014/03/23/5858ca52-8441-11e3-bbe5-6a2a3141e3a9_story.html?utm_term=.cbaf3ea5be51 (Accessed on August 16, 2016).

Kusakabe, Kyoko and Veena, N. (2015). "Gender and Land Tenure in the Context of Disaster", in *Gender and Land Tenure in the Context of Disaster in Asia*, ed. Kyoko Kusakabe, Rajendra Shrestha and N. Veena. New York: Springer International Publishing.

Limbu, Amrita and Sijapati, Bandita (2012). *Governing Labour Migration in Nepal: An Analysis of Existing Policies and Institutional Mechanisms.* Kathmandu: Himal Books.

Luitel, Nagendra P., Mark, J. D. Jordans, Adhikari, Anup, Upadhaya, Nawaraj, Hanlon, Charlotte, Lund, Crick, and Komproe, Ivan H. (2015). "Mental Health Care in Nepal: Current Situation and Challenges for Development of a District Mental Health Care Plan", *Conflict and Health*, Vol. 9, No. 3, pp. 1–11.

Maharjan Amina, Prakash, Anjal, and Gurung, Chanda Goodrich (2016). "Migration and the 2015 Gorkha Earthquake in Nepal: Effect on Rescue and Relief Processes and Lessons for the Future". *Hi-Aware.* September. Available at http://www.hi-aware.org/index.php?id=211 (Accessed on January 15, 2017).

Maharjan, Amina, Bauer, Siegfried, and Knerr, Beatrice (2012). "Male Outmigration: A Case Study in the Hills of Nepal", *Gender, Technology and Development,* Vol. 16, No. 1, pp. 95–123.

Ministry of External Affairs (2015). "India's Rescue and Relief Effort in Earthquake-hit Nepal". Media Briefing, New Delhi: Ministry of External Affairs, Government of India. Available at http://www.mea.gov.in/media-briefings.htm?dtl/25135/Media_Briefing_by_Foreign_Secretary_and_Others_on_Indias_Rescue_and_Relief_Effort_in_Earthquakehit_Nepal (Accessed on July 11, 2015).

Ministry of Finance (2015). "Expenditure Allocation for Fiscal Year 2014/15". Summary, Kathmandu: Ministry of Finance, Government of Nepal. Available at http://www.mof.gov.np/uploads/document/file/Red%20Book%20FY%202014-15_20140917050612.pdf (Accessed on June 11, 2016).

Ministry of Labour and Employment (2016). "Labour Migration for Employment from Nepal: 2014/2015". Annual Status Report, Kathmandu: Ministry of Labour and Employment, Government of Nepal,

1–114. Available at http://www.ilo.org/wcmsp5/groups/public/—-asia/—-ro-bangkok/—-ilo-kathmandu/documents/publication/wcms_500311.pdf (Accessed on August 4, 2016).

Ministry of Law, Justice & Parliamentary Affairs (2016). "Nepal Gazette, Government of Nepal". Housing Recovery & Reconstruction Platform. 13 January. Available at http://www.moudclpiu.gov.np/public/filesmanager/12.pdf (Accessed on December 9, 2016).

Naiz, Unaiza (2010). "Pakistan Floods 2010". World Psychiatric Association. October. Available at http://www.wpanet.org/detail.php?section_id=7&content_id=926 (Accessed on December 30, 2016).

National Disaster Risk Reduction Centre (2016). "Assessing the Impact of Nepal's 2015 Earthquake on Older People and Persons with Disabilities and How Gender and Ethnicity Factor into That Impact". *CBM International.* March. Available at http://www.cbm.org/article/downloads/54741/Assessing_the_impact_of_2015_nepal_earthquake.pdf (Accessed on November 25, 2016).

National Planning Commission (2015). "Nepal Earthquake 2015: Post Disaster Needs Assessment, Vol. B". Assessment Report, Government of Nepal, Kathmandu: National Planning Commission, 1–314. Available at http://www.npc.gov.np/images/category/PDNA_volume_BFinalVersion.pdf (Accessed on August 16, 2016).

National Seismological Centre (2015). "Recent Earthquakes". Department of Mines and Geology, Government of Nepal. Available at http://www.seismonepal.gov.np/index.php?action=earthquakes&show=recent&page=4 (Accessed on November 26, 2016).

Nepal Disaster Risk Reduction Portal (2016). "Updated Report of Effects due to Disaster Events". Nepal Disaster Risk Reduction Portal. Available at http://drrportal.gov.np/reports (Accessed on November 11, 2016).

New York Times (2015). "After Earthquake in Nepal, Sanitary Menstruation Practices at Risk". 1 May. Available at http://nytlive.nytimes.com/womenintheworld/2015/05/01/after-earthquake-in-nepal-sanitary-menstruation-practices-at-risk/ (Accessed on August 25, 2016).

NRA (Nepal Reconstruction Authority). (2017). "Rebuilding Nepal". Kathmandu: Nepal Reconstruction Authority, 1–12. Available at nra.gov.np/uploads/docs/84LdZ2BkQ8161026092851.pdf (Accessed on January 27, 2017).

Oxfam (2016). "Single Women and Nepal Earthquake". Oxfam. Available at https://www.oxfam.org/sites/www.oxfam.org/files/ib-i-am-alone-nepal-earthquake-single-women-230716-en.pdf (Accessed on November 25, 2016).

Pincha, Chaman (2008). *Indian Ocean Tsunami through the Gender Lens: Insights from Tamil Nadu, India.* Mumbai: Earthworm Books.

Pradhan, Tika R. (2015). "Onsari First Woman Speaker in Nepal's History". *The Himalayan Times*, October 17.

Ramdas, Kavita N. and Chew, Lin (2005). "Caught in the Storm: Impact of Natural Disasters on Women". The Global Fund for Women. Available at https://www.globalfundforwomen.org/wp-content/uploads/2006/11/disaster-report.pdf (Accessed on December 16, 2016).

Regmi, S. K., Pokhrel, A., Ojha, S. P., Pradhan, S. N., and Chapagain, G. (2004). "Nepal Mental Health Country Profile", *International Review of Psychiatry*, Vol. 16, No. ½, pp. 142–149.

Robinson, B. A. Hannah (2015). "Chaupadi: The Affliction of Menses in Nepal", *International Journal of Women's Dermatology*, Vol. 1, pp. 193–194.

Shakya, K. (2016). "Earthquake: Impact on Nepalese Economy and Women", *Lowland Technology International*, Vol. 18, No. 2, pp. 75–82.

Shaw, Rajib and Kobayashi, Masami (2001). "The Role of Schools in Creating an Earthquake-Safer Environment". Disaster Management and Educational Facilities. Greece: Prevention Web. 1–7. Available at http://www.preventionweb.net/files/5342_SesiRoleSchoolsEQSafety.pdf (Accessed on January 16, 2017).

Sherpa, Dechenla (2007). "New Vulnerabilities for Mountain Women: A Different Light on the Greater Himalaya". Women of the Mountains Conference. Utah: Utah Valley State College in Orem. p. 11. Available at http://condesan.org/mtnforum/content/new-vulnerabilities-mountain-women-different-light-greater-himalaya (Accessed on January 11, 2017).

Sijapati, B., Limbu, A., Baniya, J., Bhandari, A., Bhatarai, A., Kharel, S. and Pathak, D. (2015). Migration and Resilience Experiences from Nepal's 2015 Earthquake. Kathmandu: Centre for the Study of Labour and Mobility.

Simkhada, P., van Teijlingen, E., and Marahatta, S. B. (2015). "Mental Health Services in Nepal: Is It too Late?" *Journal of Manmohan Memorial Institute of Health Sciences*, Vol. 1, No. 4, pp. 1–2.

Solotaroff, Jennifer L. and Pande, Prabha Rohini (2014). "What Underlines Risk Against Women in South Asia?", in *Violence against Women and Girls: Lessons from South Asia*, ed. Jennifer L. Solotaroff and Prabha Rohini Pande. Washington, DC: The World Bank Group. pp. 73–135.

Stoltman, Joseph P., Lidstone, John, and Dichano, Lisa N. (2007). *International Perspective on Natural Disasters: Occurrence, Mitigation, and Consequences*. The Netherlands: Springer.

Taneja, Nisha, Samridhi, Bimal, and Dayal, Isha (2015). "Nepal's Transit through India: An Assessment", *The Economic & Political Weekly*, Vol. 7, November, pp. 18–21.

The Economist (2014). "The South Asia Women's Resilience Index". Actionaid, *The Economist*, 1–65. Available at https://www.gdnonline.org/resources/The%20South%20Asia%20Women's%20Resilience%20Index%20Dec8.pdf (Accessed on December 5, 2016).

The Indian Express (2016). "First Woman Chief Justice of Nepal, Sushila Karki, Takes Oath". 11 July. Available at http://indianexpress.com/article/world/world-news/first-woman-chief-justice-of-nepal-sushila-karki-takes-oath-2907173/ (Accessed on October 12, 2017).

The Kathmandu Post (2015a). "Embassies Ask Host Countries to Grant Nepali Workers Leave". 29 April. Available at http://kathmandupost.ekantipur.com/printedition/news/2015-04-29/embassies-ask-host-countries-to-grant-nepali-workers-leave.html (Accessed on December 11, 2016).

The Kathmandu Post (2015b). "Migrants Return for Funeral, Rebuild Life Shattered by Quake". 6 May.

The Kathmandu Post (2015c). "Women Seek Role in Reconstruction Body". 19 November. Available at http://kathmandupost.ekantipur.com/printedition/news/2015-11-19/women-seek-role-in-reconstruction-body.html (Accessed on August 16, 2016).

The Kathmandu Post (2017a). "Getting the Facts Straight". 6 January. Available at http://kathmandupost.ekantipur.com/news/2017-01-06/getting-the-facts-straight.html?fb_comment_id=1425884790755159_1426115104065461#f8a468bc3d777c (Accessed on January 8, 2017).

The Kathmandu Post (2017b). "Nepal Earthquake: Reconstruction Update". 21 January. Available at https://thehimalayantimes.com/business/post-earthquake-nepal-reconstruction-update/ (Accessed on January 29, 2017).

The United States Agency for International Development (USAID) (2015). "Nepal Earthquake". Available at https://www.usaid.gov/sites/default/files/documents/1866/nepal_eq_fs20_06-12-2015.pdf (Accessed on November 24, 2016).

Trading Economics (2016). "Nepal GDP". 6 December. Available at http://www.tradingeconomics.com/nepal/gdp (Accessed on December 6, 2016).

UN Women (2016). "Women's Leadership and Political Participation". UN Women. Available at http://www.unwomen.org/en/what-we-do/leadership-and-political-participation (Accessed on July 16, 2016).

UNDP (United Nations Development Programme). (2015). "Human Development Report 2015". United Nations Development Programme. Available at http://hdr.undp.org/sites/default/files/2015_human_development_report.pdf (Accessed on June 25, 2016).

UNDP (United Nations Development Programme). (2016). "Gender Inequality Index (GII)". Available at http://hdr.undp.org/en/content/gender-inequality-index-gii (Accessed on December 2, 2016).

UNICEF (The United Nations Children's Fund). (2015). "12 Babies Born Every Hour without Basic Healthcare in Worst Hit Areas". Press Release, Kathmandu: UNICEF. Available at https://www.unicef.org/media/media_81896.html (Accessed on December 19, 2016).

United Nations in Nepal (2016). "Strategic Summary". Kathmandu: UN Nepal Information Platform, 1–32. Available at http://un.org.np/sites/default/files/20160622_nepal_eq.pdf (Accessed on December 1, 2016).

United Nations Office for the Coordination of Humanitarian Affairs (2016). "Nepal Gender Profile". UNOCHA, 1–10. Available at http://un.org.np/sites/default/files/nepal_gender_profile_2016.pdf (Accessed on November 20, 2016).

Valdés, Helena Molin (2009). "A Gender Perspective on Disaster Risk Reduction", in *Women Gender and Disaster: Global Issues and Initiatives*, ed. P. G. Dhar Chakrabarti and Elaine Enarson. New Delhi: Sage Publications. pp. 6–18.

Vanitha, A. and Ashokan, G. V. (2016). "Disaster Response under One Health in the Aftermath of Nepal Earthquake, 2015", *Journal of Epidemiology and Global Health*, Vol. 7, No. 1, pp. 91–96.

WHO (World Health Organization) (2005). "Bulletin of the World Health Organization". WHO. 1 June. Available at http://www.who.int/bulletin/volumes/83/6/infocus0605/en/ (Accessed on December 11, 2016).

WHO (World Health Organization) (2014). "Mental Health Atlas Country Profile 2014". Annual, WHO. Available at http://www.who.int/mental_health/evidence/atlas/profiles-2014/npl.pdf (Accessed on January 19, 2017).

WHO (World Health Organization) (2015). *Humanitarian Crisis after the Nepal Earthquakes 2015: Initial Public Health Risk Assessment and Interventions.* Initial Public Health Risk Assessment, Kathmandu: WHO. pp. 1–32.

World Economic Forum (2011). *GGG Report 2010.* Geneva: World Economic Forum, 1.

World Economic Forum (2013). *GGG Report 2013.* Geneva: World Economic Forum.

World Economic Forum (2016). *GGG Report 2016.* Geneva: World Economic Forum, 1.

World Trade Organization (2016). "Country Profile". Available at http://www.who.int/countries/npl/en/ (Accessed on December 5, 2016).

8

Women's Participation in Relief Work Utilizing Social Media Network

Shanmugaraj G. and Balasaravanan T.

Introduction

Cuddalore district (in Tamil, 'Kadal' is the word for sea and 'Oor' means town) is one of the costal districts of Tamil Nadu and is situated at the end of Cauvery delta. Due to this geological proximity, the place is often affected by rain and cyclone almost every year. Also, this district has the highest number of people living in thatched huts in Tamil Nadu. One volunteer in the field says:

> Natural Calamity is not new here. For instance, the place was affected by Thane Cyclone in 2012 and many others previously. In those times, public-to-public voluntary support was not available for affected people's relief. The government machinery took action very slowly and in a piecemeal manner, so people had to mend themselves.

Cuddalore was recently affected by some major natural disasters such as the 2004 Tsunami and the Cyclone Thane of 2011. Gokhale (2008) observes that women suffer more in disasters than men. They become weak both physically and mentally. The author observes that during the 2004 tsunami rehabilitation programme, women in

Cuddalore recovered slowly with the help of NGOs. She also lists the role of women in previous disasters and their rehabilitations. The post-disaster economic empowerment of women was the solution for rebuilding their family life.

The reports state that during November–December 2015 flood, Cuddalore was severely affected by heavy rainfall and it resulted in the loss of life and property. As a result of the heavy rainfall, the houses were submerged and the systems of transportation and communication were completely disrupted. The people in these worst-hit areas faced difficulties such as short supply of essential materials including food and drinking water. Common people across the world came forward and offered aid, shelter and food. Due to the lack of work force, the public also joined hands with army in rescue work and offered help to the affected people. After relief work started in Cuddalore district, most of the helpers from various places moved to Cuddalore to help by hiding their power of identity. The flood victims had to face a lot of difficulties in getting enough food, water and other essential materials due to the lack of coordination in the distribution of relief materials. Abraham (2015) found that even though the rescue systems on social media were seemingly secular and free of gender bias, inappropriate behaviour in the form of indecent calls and texts were faced by many women. Apart from these issues, the women volunteers were successful in carrying out the relief work in Cuddalore. Women were involved more in social media coordination rather than in the work on the field due to security reasons.

The primary aim of the study is to understand the role of women volunteers in December 2015 flood relief work in Cuddalore. This study investigates how social media helped them carry out the relief mission successfully by breaking gender stereotypes. It also elaborates upon the importance of mobile communication that helped them to organize rehabilitation process, to develop their leadership quality and decision-making ability in emergency situations. In-depth interviews of five women volunteers are taken as primary resources in this chapter. Their being supportive and ready to share information made this chapter possible.

Research Method

Gee (2005: 21) defined 'Discourse' as *'the ways of combining and integrating language, actions, interactions, ways of thinking, believing, valuing, and using various symbols, tools, and objects to enact a particular socially recognizable identity*'. Based on this, we can analyse the discourse of women volunteers who mingled with men volunteers during the flood period.

The study was conducted in March 2016. By adopting convenience sampling method, five women volunteers from various places were chosen for telephone interview. The respondents are Anitha Raj (40) from Chennai, Sara Sithara (25) from Dindigul, Malarvizhi (47) from Chidambaram, Anita Sathiam (47) from Neyveli and Krithika Tharan (41) from Bengaluru, Karnataka.

Tamil Nadu flood disaster in December 2015 necessitated the women volunteer's involvement. It should be noted that this was their first volunteer activity. Among the five women, Anita Raj, Anitha Sathiam and Malarvizhi were involved in work both on (affected area) and off the field (logistic and communication points) as volunteers. Anitha Sathiam and Malarvizhi are from Cuddalore district. Anitha Raj from Chennai was involved in both on and off the field work in Chennai. She helped in Cuddalore disaster area through mobile communication. Krithika Tharan and Sara Sithara were involved in off the field voluntary work in Cuddalore disaster area. Four of the five women were Hindus and married, whereas one of them, Sara Sithara, is a Muslim and lives with her parents.

All the five women volunteers are Tamil speaking. They declared that they do not go outside their homes as much as the men in their family do. Therefore, most of their communication with the outer world is done through mobile phones and Internet. Today, mobile phones with the help of applications such as WhatsApp and Facebook provide us with plenty of opportunities. All the five volunteers have a smart phone and a laptop or computer at their homes. According to media dependency theory (DeFleur and Ball-Rokeach 1976), there is an internal link between social media, volunteers and the affected people. Here social media and mobile communication are greatly depended on volunteers, helpers and the affected people. When these volunteers found it difficult to contact their relatives, they depended on social media.

Women Breaking the Silence by Using Social Media

In India, as Nirmala (1996) observes, gender inequality is still prevalent when it comes to household chores despite the emergence and growth of several liberation movements for women. If we talk about gender stereotype through discourse, women are not treated the way men are treated. Kumar (2015) observes that there is a technological divide between men and women in terms of the use of mobile media. Patriarchal practices play a vital role in structuring the perceptions about women's interaction with mobile phones. Homemakers do not get much exposure or time to explore the wide range of technological possibilities available on mobile phones. The volunteer Anita Raj says:

> I am a housewife and I usually manage to take care of my children and family along with the household chores. I go out to drop my children at their school in the morning and to pick them in the evening. Meanwhile, I go to the market to buy vegetables and other things. Therefore, I cannot meet with my friends or make new acquaintances. The only people that I get to meet are my family members. Facebook introduced me to new faces and with the help of it I could be in touch with my old friends too.

Homemakers do not get much opportunity to have a face-to-face communication with others outside their family. Social media gives opportunity to expand their communication possibilities. One of the volunteers, Sara Sithara says,

> As I belong to the Muslim community, I had to deal with the restrictions laid down by my family in matters like going out of my home. While preparing for IAS exam, I had the opportunity to interact with some people whom I met at the study centre. After that, I had to come back to my native place due to health issues. I could not meet my friends very often because of it. However, I can communicate with them over phone, especially, through WhatsApp and Facebook. I got good friends through Facebook, even though I never personally contacted or met them.

When religion and family restrict women in communicating with the outer world, social media helped them to break it. On social media, we can find both men and women interacting with each other even if they do not know each other otherwise. Anita Sathiam had a similar point to make when she said:

> After I started using Facebook, I started developing an interest in Photography. I found a photography group where people encourage each other in their photographic work. This happened when I was taking bed rest due to health issues. And I learned photography with the help of this Facebook Photography group.

Therefore, Facebook is not just for entertainment, but also educates people and helps them empower the silenced people or groups in a society. Only a few women come forward to the public space and got involve in social activities such as politics, journalism and arts. Social media gives open space for them to express and talk freely. Many women are writing articles about politics and poems and some are posting photographs and thereby establishing their art that way. Krithika says,

> I used to write articles and poems on Facebook and because of that, I have 15,000 followers on Facebook including women and men. Without their support, I would not have thought about trying to do this risky relief work management.

We can assume that women can get plenty of followers on Facebook and they get appreciated for their work through the posts. In addition, they are encouraged when their posts are viewed and liked by others. Facebook acts as a window to see the world outside for women who are gridlocked in a limited social space.

Women's Communication Climate

With the above analysis, it can be observed that in India, due to social and cultural reasons, the freedom of communication for women is limited. However, the accessibility and affordability of social media enabled them to communicate with the public more now. Social media gives opportunity for both men and women to mingle with each other. According to Mohindra and Azhar (2012), men and women have different communication behaviours. Men will take advantage of modern communication to establish their hierarchy and supremacy. Women, on the other hand, will use more of an expressive style of communication as they are more sensitive to issues than men are. Women will be able to build, maintain and strengthen the relationship. Odine (2013) notes that Arab women are empowered in social media field by presenting women's issue to the public. Women interact with one another by creating communities on Facebook and Twitter. Arab women expanded their communication circle across their religion and culture. They came together, raising their voice through social media. In India, sixty million women are more active and dependent on Internet despite being lesser in number than men[1]. In terms of their online activity, e-mail, search and social networking are on the top list. Social media gives the space to everyone to explore their thoughts in the form of text and visual. Women's empowerment grows with the growth of communication. Like men, now women are also using all communication technologies with great ease. Traditionally, a woman wanting to talk with a non-family male for a long time is not appreciated by her people. In this kind of situation, Internet helped them to break gender stereotypes. Now women are more active in using social media network (SMN) rather than personal communication to reach others. Even though this is not a real freedom for women, they believe that they have achieved freedom of expression virtually through social media. This virtual freedom gives them a large space to express their skills. The volunteer Anita Sathiam tells, *'I used to post photographs in Facebook photography group to get appreciation'*. Facebook is very much helpful for women to establish themselves on social media as writers, photographers and journalists and to publish their work. During Tamil Nadu floods in 2015–2016, these women Facebook users utilized this emerging media and communication climate. So they took initiative to help the rescue and relief work with their Facebook contacts.

Anitha Raj says,

> I have been living in Singapore for the past 6 years after I got married from India. During that time, I missed my friends and relatives, as I could not contact them frequently. Facebook helped me be in touch with a few of my Indian friends. Just 3 months before the disaster, I came back to India with my family to get settled in Chennai. Now I do not feel that I am missing my Singapore friends. All are in contact with me through WhatsApp and Facebook. One of my Singapore friends persuaded me to start the relief work and from there she raised funds. They are in contact with me through both WhatsApp and Facebook.

It is found that social media transcends place, gender and religious and cultural barriers. Distance is not a matter on social media space. Women could connect with their friends across the country with the help of SMN. Malarvizhi says,

> Usually I don't keep more friends on Facebook. I have changed my Facebook privacy settings so that my posts can be seen only by my friends. I have only 500 friends. I do not usually accept any

> unknown male user's request on Facebook as I am afraid that it might create trouble for me. However, during the flood relief time, I changed all my privacy settings to 'public view' so as to make my posts reach everyone and to support others who were ready to take initiative for social activity.

The different communication behaviours of different genders result in women becoming hesitant to communicate with any unknown male on social media. It again limits women's freedom of communication.

Understanding the Flood Risk and Women Volunteers' Communication

Lack of communication in a situation of disaster makes relief and rehabilitation process more risky. Reuter et al. (2015) states that social media supports the volunteers in disaster management. As women are more active on social media, their role becomes vital in risk management. This chapter is important as it intends to deliver the activities of women volunteers on social media space. Four among the five volunteers whom I interviewed are married and two are homemakers and two are helping their husbands' businesses. All are very much active on social media such as Facebook and WhatsApp. Krithika says,

> I read the news about Chennai getting heavy rainfall on Facebook. After that, I contacted my friends in Chennai through WhatsApp and Phone calls to know their status. Initially, when heavy rain starts, there is no communication problem. After continuous raining of two, three days, my Chennai friends were disconnected. I used to get message from them daily saying there is heavy rain. Suddenly they were unreachable. After that, I watched news on television that Chennai was flooded. Then on Facebook, people were continuously posting about the status of Chennai. So, I tried to communicate with my friends anyway.

The communication network failure made Krithika understand the risk even though she was far away from the flooded area. The other volunteer Sara says:

> As I am preparing for IAS in Chennai, I have friends in Chennai. Before flood, my health was not good due to continuous rain. Therefore, I came back to my native town Dindigul where my parents are staying, to get bed rest and follow the medication.

> However, I was in touch with my friends through WhatsApp group. Based on that, I knew the status of Chennai up to date. My uncle was staying with my grandfather in Chennai where I stayed. When all means of communication got disconnected, I was scared.

It is found that communication is important, but sometimes not communicating is also important, particularly, during the disaster to avoid choking the communication network. Anitha raj says,

> When heavy rain started, the schools announced holidays, and we were unable go out to get groceries to cook food. My husband could not go for his job. Somehow, he managed to get groceries from nearby shops. After the disconnection of all means of communication and electricity, we could not go out of our home, because our area got flooded. In this situation, we managed with available food, whatever we had at home. There was neither signal nor battery charge on our phones for about two days. Then gradually electric supply got restored. All mobile phone networks were dead. Luckily, I was using BSNL landline internet connection, and it was working. Therefore, I contacted my kith and kin. I informed them that I was safe.

Men know about the basic needs of a family such as food and groceries. So they help women in family rehabilitation when there is a risk.

Krithika recalls:

> We were initially giving food, then we started supplying relief materials such as cloths, scarf and the other things to the affected people. One of my friends, a woman, informed me that they were desperately in need of napkins. When she was on the Chennai disaster field, girls were asking for Napkins. So we posted about the requirement of napkins. So people started to purchase napkins that were distributed for women in the affected area. Some of my friends asked me if the village people would know about napkins and if they would use it. I told them that if they do not know we will teach them about it.

Women knew the basic needs to run a family during rehabilitation and in their interaction, they could understand more about women's vulnerability and health issues, which men could not understand that well.

Available Technological Support for Women Volunteers to Handle the Emergency

In any case of emergency, communication is the most important aspect that helps to understand the present and evolving risk. Covello (1992: 359) describes risk communication as the *'process of exchanging information among affected people and relief materials about the nature (of disaster), magnitude (of wreck), significance (of aid), and control of a risk'*. One who is affected by or the one who witnessed the situation in the field can explain all these things.

Women are highly vulnerable than men after the disaster (Gokhale 2008). In disaster management, women have to take personal care while taking care of her family and society (Gordon 2013). Social capital offers for women who are affected by disasters are important because it would help overcome the unique vulnerabilities of women and strengthen their capabilities in preparing for, responding to, and recovering from disasters (Emel Ganapati 2012).

Anita Sathiam notes:

> Photography is my hobby. I took photographs of the disaster and then shared it in Facebook photography group. When I went to take photos, people from the affected area requested me to give food. I got emotional when they told me that they have been starving for the last two days. On the field, I got acquainted with a young boy and his friends from the village who were actively helping people. With their support, I did some small help with the money I have. Similarly, seeing my photographs, some Facebook friends contacted me to help them. Then I investigated on the field about the needs of the affected people and reported about it to the helpers.

This is how observing the situation in the field changes a person to involve in voluntary work. As she is from a nearby place, Anita Sathiam sought the help of local people who knew about their village very well. Photographs taken from the flooded area helped increase the interaction with the public.

Sara says:

> I found that Tamil Television Channels were focusing on Chennai flood and giving the report on Chennai. However, National Television channels were not giving clear reports about Chennai. The people who know their friends, who were affected by the flood in Chennai, continuously posted about these things on social media.

So, I, with my friends who are in Chennai, wrote about the Chennai flood seeking help.

From the field of disaster, the amount of risk was measured and it was informed to the rescue and relief teams in order to organize relief materials. Due to the lack of information network, and the confusion created through social media, the affected people and their relatives depended on the local helpers to get immediate relief.

Malarvizhi explains:

> As I support my husband in the administration of his hospital and petrol bunk, I meet different local people daily. Our employees from the affected area used to tell me about the situation. In addition, I started helping the affected people with our employees' and my family's support. As I have a professional camera, I took some pictures when I went to help people and posted it on Facebook. My friends who know me through Facebook wanted to help those people. Therefore, they collected money from various places and asked me about what to do. I asked them to send the money to me if they had trusted me, so that I can buy relief materials and distribute it to the affected people. They started collecting money and sending me. Like that, I received more fund from Facebook contacts. With that money, I bought relief materials according to the need of the affected people. I bought groceries like rice, dhal, cooking oil, wheat and other things needed for cooking. I also bought some biscuits, water bottles, candles, matchboxes, mattress, blankets, etc., that the affected people needed. I used to go by my own car with my employees to get relief materials. They supported me in packing and distributing the materials in the affected area. My husband also used to come with me after his duty hours, during the night to lend a helping hand in the relief work. I also directed some people, who contacted me through Facebook, to the affected area.

The women, who used to take and publish photos from the affected area, were greatly appreciated and depended on for reliable information.

Women's Reliability in Relief Management with the Help of Mobile Communication

In emergency situations, photographs are viewed as sources for information rather than pieces of art. Malarvizhi says,

> I shared photos and posts regarding Chennai and Cuddalore flood on Facebook. Some of my Facebook friends, who knew that I was from Cuddalore district, contacted me through Facebook to get the route to reach the flood-affected place. And so an unknown person from Kuwait, who was not even my friend on Facebook, also contacted me through Facebook to help the people.

People from various places wanted to know about the affected area to help them. Affected people and their relatives wanted to communicate with each other even after the disconnection of all communication. Outside helpers wanted to get the right information from the right people. Therefore, they even depended on unknown volunteers through reliable social media.

They did not think about whether they could do or not. They thought that they should do something supportive for the relief work. Therefore, they fully dedicated themselves as volunteers. They wanted to help the affected people with their knowledge as much as possible. Therefore, they believed in the kind of support that social media can provide with. Sara recalls:

> After seeing many posts regarding Cuddalore which indicated that 'Cuddalore was affected more but none of the media and public was doing anything for rescue and relief work', I discussed with my Facebook friends Manushi Bharathi and Isai Prakash and decided that we would do relief work for Cuddalore district. Manushi Bharathi's home is at Puducherry which is near to Cuddalore. We planned to set up a control room to connect with the volunteers. First, I created a WhatsApp group, added my 20 friends as admins and then, they added their friends all over the world to that group. Local friends started going to Cuddalore to do relief work on the field. Some friends, who were far away from Cuddalore, raised funds and arranged relief materials. Some friends from foreign countries, who really wanted to help, used to recharge our mobile phones that were used for the work in the control room. Therefore, we could contact anyone at any time without hesitation. At the initial stage, we faced a lot of difficulty in getting electricity in the flooded area. Therefore, we asked volunteers to use basic model mobile phones, which have long battery life, so that they can give reports on the affected area and its people's requirement through calls and text messages. People in the control room collected the information from the volunteers, and then we shared it to the public through social media network. In our posts, we gave our

> contact numbers in order for the affected people, helpers and the unknown volunteers to contact us immediately. We received more than 500 calls a day from both the donors and the needy during the relief work. We could sleep only 2 to 3 hours a day when we got more resources to distribute.

They believed they could do the relief work as much as possible with the available work force, charity source and non-material resources.

Problems Handled by Women Volunteers with Recognizing their Identity

'Whether the difference in the communication behaviours of men and women matter in flood relief management?' is an important question. What happens when there is a need to communicate with unknown people on social media? Sara answers:

> Sometimes unknown boys used to send messages as they wanted to help. When I responded to them, some of them talked in an indecent manner. I did not give up or stop giving my number just because I came across with some such people. At the same time, we contacted more unknown people, whom we found on Facebook, in order to get help.

The author, being a man, did not face any harassment when he worked as a volunteer. At the same time women volunteers faced this kind of problems when they shared their contact numbers on social media.

Anitha Sathiam says:

> When I did not have enough resources for food, I had to contact an unknown person who once posted on Facebook thus; 'If you need help, we are providing food, water, so contact on this number ...' I believed it and made those people wait for that help. But the food did not reach them. They said that some people on the way blocked their vehicle and so they gave it there. I felt very bad because those people stayed hungry because of me. After that, I decided not to contact any unknown person. If any unknown person contacted me with relief materials, I would direct them to reach the right place. When known Facebook friends contacted me to help the affected people, I would ask them to send the money to me if they had trusted me, so that I can do all the arrangements myself.

When a volunteer faces some sort of misbehaviour from an unknown person, it scares people to approach unknown people. She also faced some other problem which she revealed:

> when I went to the field alone, some people stopped my vehicle on the way and said, 'you should give relief material here. Don't give anything to the SC people'. In order to manage that situation, I called known local youngsters and asked them to help me in distributing relief materials. After that, I did not go to the field very often. If there is a vehicle going to Cuddalore, I would connect with the local volunteers; they would come to their border with four or five youngsters as security guards; then they would take the vehicle to their place and after the distribution of relief materials, they would call me to give report.

Even after being threatened by the local people, Anitha Sathiam continued her relief work from home with the help of native volunteers by using mobile communication.

Krithika recalls:

> Once I phoned one of my relatives in Cuddalore to give the relief material to an unknown affected place. However, he told me; 'it's Dalit area. So, you do not go there', and refused to help me. Then I approached another volunteer group to reach the place. After that, I started writing about the risks and issues in distributing relief materials in Dalit area on Facebook. I received abusive comments from some unknown people for those posts. Occasionally, I deactivated my account for a few days. Then I continued relief work with the existing mobile contacts.

When the public finds women surpassing the unjust social barriers, they try to mute them by abusing them. Women face a lot of hurdles when they try to use their power through social media in a disaster situation like this.

Discussion and Conclusion

During the December 2015 flood, the author worked as a volunteer to distribute relief material to the affected people in Cuddalore district, Tamil Nadu. The author has worked with both men and women volunteers; most of whom were unknown to him until that moment. In this chapter, the author intends to focus particularly on the women relief workers that he came across with. The method of in-depth interviewing is employed for this research in order to understand how they utilized the media and how effectively they communicated, organized and networked the relief work. Social media has played a vital role in facilitating women's participation in the natural calamity relief work in Cuddalore district. Twitter, Facebook and WhatsApp are the media that they have utilized as immediate communication tools. Due to the lack of security, women volunteers worked mostly during the day time. However, they utilized the information and communication technology more on the field and also after coming back home from the flood-affected area in order to coordinate the work. Mohindra and Azhar (2012) carried out a research on communication approaches of men and women in a working field. They found both are having different types of communication in working place. The research question is if these two different genders work in a disaster field, how communication process will happen in emergency situation.

In an emergency, male dominance persists even though individual identities are not important. People from all religions and communities wanted to join hands to help each other in the turbulent time of flood[2]. Social media helped them in crossing their cultural limitations in terms of mingling with others on the disaster field. Economic support is needed for a family affected by a disaster in order to recover from it and reset itself. They need economic support to get their basic needs such as food, dress and house. Gordon (2013) suggests that woman leads in rehabilitation of her family as well as in helping the society in terms of preparing food in emergency. Considering gender, woman should be given more importance rather than man in recovering from the disaster-affected situation. She has to recover her health, mind and her economic status. Knowing women's status in the disaster-affected family is very important. They should get space to express their needs. Another important observation from the field is that in the rural areas of Cuddalore District, women follow their culture. The lack of media literacy and women not getting free access to social media are an issue. The volunteers on the field who access social media were mostly men. Women were reluctant to talk about their needs to men volunteers directly, and so they asked through their male family members who would communicate with the male volunteers. The author found a significant lack of women volunteers on the field. The author had worked along with two women social media volunteer coordinators. It is understood that women on the field had to face a lot of threat and violence because of their gender and caste. However, social media enabled women to display their power. They focused on getting help from more people, be they are known or unknown. Normally, people cannot share their contact details on public platforms such as Facebook. During emergency, their contacts were shared through Facebook to everyone. Some Facebook users contacted the women volunteers and offered to help them. Female volunteers also contacted known and unknown people through Facebook in order to offer and get help. Even though this process made women transcend traditional gender boundaries, they were constantly reminded of the stereotypical gender roles imposed upon them by the society. When they wanted to talk about the political discourse on social media, they were suppressed by the patriarchal society. They were humiliated when they wanted to be social with unknown men. However, social media gave the power to overcome some of these issues while participating in the relief work during emergency. With the support of their friends and family, they were able to get involved in flood relief work both on the field and off the field. Unmarried women and homemakers were engaged in relief work. Despite having their own family responsibilities, each volunteer wanted to help. As women are more sensitive, they could feel the pain of others who were not able to contact their kith and kin when all means of communication got disconnected. So, they became volunteers in order to collect, coordinate and distribute the needs of the affected people.

Sukul (2014) observes that the usage of mobile phones differs between urban and suburban youngsters. The rural Cuddalore youths are not well-versed with the technology and the usage of mobile phones. So, rural people had to seek the help of others who knew about social media. Both men and women volunteers from urban places went to the field. They observed the situation and mediated with the help of SMN. Matanhelia (2010) illustrates that there is difference in their usage between women and men. These different people combined their mobile communication skills during Tamil Nadu floods 2015. As analysing discourse, women were revealing their gender identity in flood relief work when unknown men in the disaster field and even in social media attacked them. However, women volunteers supported the women in the affected area by providing them with sanitary napkins and other things. Most of the male volunteers (including the author) in the field distributed the napkins to the affected women. They aided the women in giving other medicines, food, dresses and groceries. Both men and women volunteers worked together in fulfilling the requirements of both men and women irrespective of their gender. Women's participation was very important in dealing with the requirements of the women in the affected area, especially as they were not hesitant to distribute and get the napkins from both men and women. All five respondents have Tamil and English communication skills. Their photographs became reliable sources of information. They were exchanging information with their skills. The flood relief work made women share more information on the social media and thereby expanded their communication circle.

Indian women feel very free to express and share their thoughts with the public through social media rather than through interpersonal communication. Social media could be considered as a weapon for women's mass communication because it enables them to communicate with the world outside their homes while assuring safety with its privacy settings. It helps people from all backgrounds, irrespective of their gender, religion, caste, culture and location, to come together. Mobile phone communication was used as one of the safest tools by women volunteers to carry out the relief mission effectively. In future, during an emergency, there should be only two segregations: the affected and the helpers.

Notes

1. "Sixty Million Indian Women Use Internet: Google Study" (20 June 2013). *Firstpost*. Retrieved from http://tech.firstpost.com/news-analysis/sixty-million-indian-women-use-internet-google-study-214240.html
2. "Chennai Floods Bring out a Supreme Show of Religious Unity", December 08, 2015. Indiaglitz. Retrieved from http://www.indiaglitz.com/hindu-muslims-christians-sikhs-and-jains-unite-in-chennai-rain-relief-operations-tamil-news-148189.html (Retrieved on 2016, January 12)

References

Abraham, B. (12 December 2015). "Shocking! Women Volunteers Involved in Chennai Flood Relief Face Harassment". *India Times*. Retrieved from http://www.indiatimes.com/news/india/shocking-women-volunteers-involved-in-chennai-flood-relief-face-harassment-248312.html

Covello, V. T. (1992). "Risk Communication: An Emerging Area of Health Communication Research", in *Communication Yearbook 15*, ed. S. A. Deetz. Newbury Park, CA: Sage Publications. pp. 359–373.

DeFleur, M. L. and Ball-Rokeach, S. J. (1976). *Theories of Mass Communication*. Enskede: TPB.

Emel Ganapati, N. (2012). "In Good Company: Why Social Capital Matters for Women during Disaster Recovery", Public Administration Review. *The American Society for Public Administration*, Vol. 72, No. 3, pp. 419–427. doi: 10.111/j.1540-6210.2012.02526.x

Gee, J. P. (2005). *An Introduction to Discourse Analysis: Theory and Method* (2nd ed.). New York, NY: Routledge.

Gokhale, V. (October 2008). "Role of Women in Disaster Management: An Analytical Study with Reference to Indian Society". The 14th World Conference on Earthquake Engineering. Beijing, China. Retrieved from http://www.iitk.ac.in/nicee/wcee/article/14_10-0049.PDF.

Gordon, L. (2013). "Preserving Family and Community: Women's Voices from the Christchurch Earthquakes", *Disaster Prevention and Management*, Vol. 22, No. 5, pp. 415–424. http://dx.doi.org/10.1108/DPM-10-2013-0175

Kumar, N. (2015). "The Gender-Technology Divide or Perceptions of Non-use?", *First Monday*, Vol. 20, No. 11. http://dx.doi.org/10.5210/fm.v20i11.6300

Matanhelia, P. (2010). "Mobile Phone Use by Young Adults in India: A Case Study". Doctoral Thesis, University of Maryland, College Park, USA. Retrieved from http://drum.lib.umd.edu/bitstream/handle/1903/10255/Matanhelia_umd_0117E_11079.pdf;jsessionid=962E9891204B5FF25901D50A924AC1F5?sequence=1

Mohindra, V. and Azhar, S. (2012). "Gender Communication: A Comparative Analysis of Communicational Approaches of Men and Women at Workplaces", *IOSR Journal of Humanities and Social Science*, Vol. 2, No. 1, pp. 18–27. doi:10.9790/0837-0211827

Nirmala, K. (1996). "Concept of a Liberated Woman Feminism in Indian Context". Doctoral Thesis, Mahatma Gandhi University, Kerala, India. Retrieved from http://shodhganga.inflibnet.ac.in/handle/10603/269

Reuter, C., Ludwig, T., Kaufhold, M.-A., and Pipek, V. (2015). "XHELP: Design of a Cross-Platform Social-Media Application to Support Volunteer Moderators in Disasters", *Disasters & Humanitarian Events* (4094–4102). ACM. http://dx.doi.org/10.1145/2702123.2702171

Sukul, A. (2014). "Gender Differentials in Mobile Phone Use: A Study of Trends among Urban and Suburban Youth". Doctoral Thesis, University of Kalyani, West Bengal, India. Retrieved from http://shodhganga.inflibnet.ac.in/bitstream/10603/61900/3/03_abstract.pdf

9

Understanding Women's Experience of Trauma and Posttraumatic Growth

A Review Study

Alphonsa Jose K. and Surendra Kumar Sia

Introduction

Disasters are sudden events that may surpass the human capacity to withstand without availing help from others. Natural disasters such as floods, storms, earthquakes, cloudbursts and tsunamis most of the time hit the human race as well as the living beings hard and the instances become traumatic when the capacity of the victims exceeds. The consequences and effects of disaster upon individual are dependent on to his or her affiliated social system (Blaikie et al. 1994; Quarantelli 1994). Among those social factors, gender, ethnicity, class, age and disability determine the degree of one's vulnerability (Ariyabandu and Wickramasinghe 2005; Blaikie et al. 1994).

Physical and psychological consequences of a large-scale natural disaster include death toll, material loss, economical burden, loss of safe environments, loss of infrastructure, Post-traumatic stress disorder (PTSD), depression, anxiety and posttraumatic growth (PTG). Even though disasters will hit everyone without any discrimination, the society and its rules will make the experience tougher or gentle. Here this study focuses upon the gendered perspective of consequences of natural disaster, with special consideration to PTG. Other known consequences of gendered society and natural trauma combination include a gender breakdown due to high death toll of females (Felten-Biermann 2006; Rees et al. 2005). Re-traumatization in the rescue camps, at times even in the forms of rapes or sexual and non-sexual assaults, forced marriage and following sudden changes in the social roles and many others will add on to the traumatic experiences of female victims. Earlier, the scientific fields considered the negative outcomes of trauma experiences widely. But recently in psychology, a new trend is observed which focuses on the positive outcome of trauma experiences: PTG. PTG explains the salutogenic outcome of trauma, while the previous trend studies the pathogenic consequences such as PTSD, depression and anxiety (Gerrish et al. 2009). Experience of positive change makes one to be more compassionate (Talbot 1999), prepares for a fully functional life (Miles and Crandall 1983), to become wiser and more accepting of life's paradoxes (Calhoun and Tedeschi 1998), with maturity and enhanced self-esteem as well, makes one more patient, tolerant, empathic and courageous (Affleck and Tennen 1991), and also makes one more spiritual and religious (Edmonds and Hooker 1992) and to have heightened existential awareness (Yalom and Lieberman 1991).

The double disadvantage of females during natural disasters appears to be contributing to PTG outcomes, this chapter further discusses the importance, dimensions and possible reasons of gender difference in PTG, with reference to varying sources of psychological traumas.

PTG: Salutogenic outcome of trauma

PTG is the construct describing the positive outcomes experienced in response to traumatic or challenging circumstances of one's life. It results from one's psychocognitive efforts to deal with the overwhelming situations or experiences (Calhoun and Tedeschi 2001). According to Calhoun and Tedeschi (2004) 'it is the individual's struggle with the new reality in the aftermath of trauma that is crucial in determining the extent to which Posttraumatic growth occurs' (p. 5). One should be clear with the fact that there is nothing inherent good in traumatic experiences, but the process of struggling with hardship arise in response to the traumatic event may push the individual to a heightened level of functioning (Linley and Joseph 2004; Malhotra and Chebiyan 2016). These approaches to trauma growth and development indicate that finding meaning in trauma results in a change in harmony, altruism, resourcefulness, spiritual development and integration of a person (Decker 1993). But these positive outcomes do not exist in pure isolation, people who reported growth following traumatic events also report negative effects such as PTSD (Malhotra and Chebiyan 2016). In summary, positive outcomes experienced aftermath a trauma coexist with the negative experience of trauma (Solomon and Dekel 2007) expanding on the relationship between trauma exposure and growth, research has examined the relationship between PTSD and PTG. For example, the relationship of PTSD and growth has been reported to be moderately positive (Maercker and Langner 2001), not significant or negative (Lehman et al. 1993); studies also reported both linear and quadratic relationship between them (Solomon and Dekel 2007). This indicates that the relationship between PTSD and growth may be unclear or positive.

A proliferation in the studies exploring growth or positive outcomes after the trauma experience can be seen after 1980s. This period witnessed the work by Calhoun and Tedeschi, and they named these outcomes as 'Posttraumatic growth'. Later a number of studies reported a similar result and as a result of increased concentration to the area, a handful of measures are available to measure PTG. These include Benefit Finding Scale (Phipps et al. 2007), Perceived Benefit Scale (McMillen and Fisher 1998), Changes in Outlook Questionnaire (Joseph et al. 2005), Posttraumatic Growth Inventory (PTGI; Tedeschi and Calhoun 1996) and Stress-Related Growth Scale (Park et al. 1996) as well as open-ended measures (Bower et al. 1998).

The pioneers in PTG study, Calhoun and Tedeschi, proposed that the new growth occurs mainly in five dimensions of one's life. Initially they focused on a three-dimensional result, however later factor analytical studies evidenced a five-dimensional structure (Taku et al. 2008). In order to identify growth, it is not necessary to experience growth in all these five dimensions, but a minimum of one dimension is enough (Tedeschi and Calhoun 1996). These dimensions include attitude towards life, interpersonal relationship, personal strength, spiritual development and new possibilities in life. The given measure of PTGI also measures the changes in these five domains aftermath the trauma (Tedeschi and Calhoun 1996). A considerable number of studies on PTG have utilized this scale and, therefore, validated the five-dimensional construct of PTG and also cross-culturally validated the phenomenon (Taku et al. 2009). The given five dimensions of PTG are as follows:

Attitude towards Life

Once the individual encounters the reality of volatile nature and vulnerability of one's life in close, the unshaken beliefs are demolished and they are forced to reconstruct these shaken beliefs with a better belief system. This cognitive reconstruction ensures individual vulnerability and difficulty in controlling and predicting certain events' in one's own life (Calhoun and Tedeschi 2001); and there occurs a shift in the priorities and importance given to certain events and some new priorities are set which were previously neglected. This contribute to a changed attitude towards life (Lindstrom et al. 2013; Ramos and Leal 2013).

Interpersonal Relationships

As widely perceived, the misery loves company. The experiences of a traumatic event make people turn towards the people around them for social support. Reflexive thinking involved in the process contribute to meaningful, warmer and intimate relationships (Tomm 1987). This increased need for support and self-disclosure may contribute to increased emotional connectedness and intimacy in relationships. Sometimes these experiences help the people to initiate or to strengthen an intimate relationship, and at times to redefine the relationships in one's life (Tedeschi and Calhoun 1996; Calhoun and Tedeschi,2004; Ramos and Leal 2013).

Perception of Personal Strength

The reconstruction of the shaken fields with stronger raw materials is observed in case of PTG too. Person becomes more prepared with recognized capabilities to deal with future challenges and adversities. This perception is not unrealistic without knowing the individual vulnerability (Calhoun and Tedeschi 2004; Lindstrom et al. 2013; Ramos and Leal 2013).

Spirituality

There is an observed change in the spiritual domain of survivors of traumatic experience. Heightened belief in a higher entity may be a way of coping in the face of trauma, and it acts as powerful resource providing comfort, peace and resolution (Kelly 2004). The meaning of suffering is very defined and rewarding according to many religions, and this may contribute to one's spiritual advancement aftermath trauma. Spiritual considerations may even ensure a causal attribution or meaning to otherwise uncertain trauma experience (Calhoun and Tedeschi 2004; Lindstrom et al. 2013; Ramos and Leal 2013).

New Possibilities

The traumatic experience opens new paths and possibilities to the survivor, through a new philosophy of life (Calhoun and Tedeschi 2004; Lindstrom et al. 2013; Ramos and Leal 2013).

Clinical utility of PTG

As we have seen in the previous section, the growth outcomes do not exist in isolation, it coexists with negative outcomes, however, the benefits of PTG is found to outshine the negative changes at almost all the times (Calhoun and Tedeschi, 2004; Malhotra and Chebiyan 2016). Even though few studies have reported a positive or nonlinear relationship between PTSD and PTG, no studies have ever reported a positive relationship between PTG and depression. Most of the studies reported either unclear or negative relationship between PTG and depression (Zoellner and Maercker 2006). Aldwin et al. (1994) found that war veterans who reported growth were less likely to suffer from PTSD symptoms later in their life. Zoellner and Maercker (2006) reviewed that few longitudinal studies reported PTG as a predictor in future reduction of distress, PTG may act as a buffer against distress following a high trauma exposure. Certain longitudinal studies provided evidence for the buffering effect of PTG as a result of interaction with time or trauma severity (Frazier et al. 2001; McMillen and Fisher 1998). It also revealed that psychological distress increased over time among people who did not experienced a growth and decreased over time among people who reported a growth (Zoellner and Maercker 2006).

Benefits resulting from PTG are not just limited to psychological health, but there are also many studies reporting a physiological benefits following trauma. Cancer survivors, who reported a PTG experience, also reported longer survival and better recovery from chemotherapy (Dunigan et al. 2007). Affleck et al. (1987) in a longitudinal study found that cardiac patients who reported a growth experience immediately after the heart attack showed a decreased level of morbidity even after 8 years. Reports of PTG are found to be associated to higher levels of self-esteem, lower levels of relationship anxiety, more comfort on depending on others (McMillen et al. 1995), lesser levels of depression among sexual assault victims (Frazier et al. 2001). A similar result of decreased level of depression was observed among stroke victims in a study conducted by Thompson (1991), it also reported an increase in meaningfulness of life. These reports revealed the adaptive nature and utility of PTG in clinical and non-clinical populations, and also provide the rationale for the burgeoning number of studies on PTG.

Process of PTG

Increases in the studies were helpful to identify certain predictors and positive associates of PTG as well as a handful of models and theories. Cognitive adaptation theory proposed by Taylor (1983) considers individuals as the active agents with the capacity to bounce back to psychological equilibrium after a trauma experience. Janoff-Bulman's (2010) shattered assumptions theory of PTSD sets the path for a glorious theoretical model of PTG in the history of growth studies. Shattered assumptions theory, as it is indicated in the name itself points that traumatic events too big to handle by the person, shatters one's basic assumptions of themselves and the world, resulting in the world which is quite unmanageable and meaningless (Janoff-Bulman 2010). PTG model was based primarily on this belief that traumatic experiences shatter one's existing schema and seismic nature initiates a new process of cognitive rebuilding and restructuring through the suffering, which basically include rumination related to stressful event. It will result in the formation of new adaptive and deliberate schema, which paves the base for PTG (Tedeschi and Calhoun 1995). Theory furthered to mention the role played by certain proximate social and cultural factors such as interpersonal interactions and social support (Calhoun and Tedeschi 1998). Theory also places an importance on cognitive and emotional processing of trauma (Calhoun and Tedeschi 2004), many distal factors such as pretrauma person, self-disclosure, schema, belief and goals also play a role along with the already mentioned proximal predictors as schema change and distress (Maercker and Zoellner 2004).

Person-centred perspective integrated with positive psychology suggests an organismic valuing theory (Joseph and Linley 2005) to explain the process of growth following adversities. Theory based itself on the person-centred perspective of organismic valuing process (OVP) which shows that the individuals possess the innate ability to know the essential and important things to them for a fulfilling life, they are intrinsically motivated to move towards beneficial goals (Rogers 1951). Sheldon et al. (2003) empirically evidenced this notion. The organismic valuing theory explicates how the trauma-related information is processed. If the traumatic event is powerful enough to disconfirm the existing assumptions, it results in assimilation of either trauma-related information into the existing assumptive world or the process of accommodation may occur. Trauma-related information may be accommodated either in positive or in negative dimensions, either way results a new assumptive world. In other words, OVP theory places the accommodation process at the central stage in case of growth (Joseph 2009; Joseph and Linley 2005). OVP theory finds its implication primarily in the support service by suggesting the need of empathic reflection of the clients' mentioned benefits after the trauma in order to facilitate growth during the client-centred therapy (Joseph 2009).

Affective cognitive processing theory of growth is also consistent with the individual's intrinsic motivation to move towards growthful outcomes. Further, a non-medicalized framework put forward a cycle of cognitive appraisal, emotional state and coping as person attempts to reunite posttrauma assumptions with trauma-related information. The theory centres on the personality, social–environmental dimensions and other role of cognitive aspects in the PTG experience. Theory marked its significance by providing a comparatively holistic process indicating the involvement of pre-existing assumptions, experienced challenges to assumptions and the inconsistency between these two acting as the trigger to the affective cognitive processing. Model also explains that the presence of intrusive thoughts, rumination and/or reflective pondering (together constitute cognitive appraisal) changes the original negative affective states to positive ones (affective states), which further contribute to the use of varying coping strategies. Assimilative or accommodative processes are utilized for the resolution of discrepancies (Joseph et al. 2012).

Important predictors of PTG

Theoretical frameworks propose certain important predictors of PTG, further information about the positive predictors and associates of PTG can be obtained from the empirical and meta-analytical study findings. Next session deals with these particularly important contributors of PTG. As we have seen above, major associates of PTG are social support, optimism, coping, rumination, religiosity, enduring distress, self-disclosure, positive affectivity and cognitive appraisal.

Social support system in the forms of friends, family and significant others act as a buffer against stress. Schaefer and Moos (1992), OVP theory, and Calhoun and Tedeschi (2004) consider social support as a key environmental resource. Many studies have reported the positive effect of perceived support resulting in better adjustment to traumatic event (Halcomb et al. 2005; Keane et al. 2006; McIntosh et al. 1993). A study conducted among Kashmir conflict victims yielded a congruent result with the finding that the perceived social support was positively associated with PTG (Bhat and Rangaiah 2015). Individuals are more likely to experience PTG, if they share their experience with others and seek tangible support from people around (Tedeschi and Calhoun 1996, 2004, 2006; Calhoun and Tedeschi 2004).

As in the case of social support, religiosity and spirituality were also strongly related to PTG (Cadell et al. 2014). They are found to be strongly related to making meaning and understanding questions of life, which further helps in coping with traumatic events and foster resilience (Peres et al. 2007) since the meaning 'making of traumatic events' is important in the PTG process, and religiosity and spirituality are found to be positively associated to PTG. Resilience is also found to be associated positively with growth and negatively with PTSD (Bensimon 2012).

A longitudinal study conducted among cancer patients proved the role of perceived social support (the autonomy supportive style of caregiving), active and planning coping strategies, and positive reinterpretation as positive correlates of reported growth. Study also revealed that autonomy support style of caregiving and problem focused coping was more reliable predictors (Scrignaro et al. 2011). Another study which utilized a structural equation modelling reported that deliberate rumination of benefits and social support was associated to PTG (Morris and Shakespeare-Finch 2011). Taku et al. (2009) also found that deliberate and constructive rumination immediately followed the trauma, and the intrusive rumination were associated to PTG. Cross-sectional study conducted among breast cancer patients revealed that PTG was in positive association with greater perceived life threat, more talking about breast cancer, higher income and longer time since diagnosis (Widows et al. 2005). Pietruch and Jobson (2012) proved that disclosure of experiences of psychosis supported the process of recovery and PTG among a clinical population experienced psychosis.

As we have seen many studies have come across a number of factors positively associated to PTG, along with all these factors, another demographic variable requires an important position, namely gender, the following section deals in breadth about the gender difference observed in the PTG experiences and discusses the various underlying factors involved in this difference.

Gender Difference in PTG

Many factors were identified and many studies are in progress to identify the processes behind the development of PTG. Among many other factors, a variable requires special mention, that of gender. Tedeschi and Calhoun (1996) evidenced that traumatized women perceived growth twice as high as those of traumatized men in a college sample using self-reported measure of PTGI. Other studies have confirmed similar results (e.g., 1996Individual differences in reports of positive changes and benefits have been found to be associated with a number of demographic and clinical variables, with higher levels of positive changes and benefits related to the female gender (Curbow et al. 1993; Tempelaar et al. 1989; Widows et al. 2005). Among these studies, Park et al. (1996) used a different self-reported measure, stress-related growth scale, in college sample to repeat the gender difference. Similar result was obtained across multiple types of trauma including cancer, HIV/AIDS, terrorism and natural disasters (Bellizzi 2004; Jang 2005; Milam et al. 2005). This results show that it is not the traumatic event, but the other factors are important in predicting the gender difference in PTG, and emphasizes the need to understand it. It is important to mention that there were also many studies which failed to report a significant gender difference (Ho et al. 2004; Polatinsky and Esprey 2000; Widows et al.

2005) and a few reported that the males experienced higher levels of PTG than women (Hooper 2003). Diversity observed in the result provoked a few meta-analytic studies as well. Helgeson et al. (2006) reported a slightly greater PTG among women ($\pi = 0.08, p < 0.001$) in their meta-analytic study which examined gender along with other important variables contributing to PTG. Another meta-analysis with a total sample of 16,076 participants from 70 studies has gone through various traumatic events including cancer, bereavement, terrorism and natural disaster, which also reported a significant, small to moderate gender difference in PTG (Vishnevsky et al. 2010).

Possible Causes of Gender Difference

This substantial variability and the clinical utility of PTG together sets the need to understand the possible reasons behind this gender difference. The present study aims to review the available researches and to discuss the major reasons behind this observed gender difference. The following section discusses the possible reasons behind the observed gender difference.

Age–Gender Interaction

Age is found as an important factor in the experiences of PTG. Specifically it is apt to say that age– gender interaction effects are more contributive in nature than simply age. Matured females are found to report higher PTG, gender differences increased with age (Arpawong et al. 2015; Vishnevsky et al. 2010). As women mature, the perception of severity or seismic nature of traumatic events is increased. Peri-traumatic distress and reported distress associated with life events are found to mediate the relationship between PTG and gender (Mooney et al. 2012).

Prati and Pietrantoni (2009) conducted a meta-analysis of 103 studies of PTG and concluded that optimism, social support, spirituality, acceptance coping, reappraisal coping, religious coping and social support seeking coping were associated with PTG among adults, and that age and gender were significant moderators of the coping–PTG relation, such that religious coping was more beneficial for older individuals and women.

Distress

Women's psychological and biological response to traumatic events are comparatively acute than that of men, including intense fear, avoidance, panic and anxiety; they are more prone to perceive an event as threatening, more stressful and with decreased personal control (Olff et al. 2007). Tolin and Foa (2006) suggested that it is not the greater rate of exposure to trauma, but the gender difference in cognitive and affective processing of traumatic events was contributing to higher perception of threat. Grief also differed significant with gender, with women showing more grief than men (Patrick and Henrie 2016). The female gender's inclination towards experiencing more growth may be attributed to their greater experience of stress as a response to stressful life event (Anderson and Manuel 1994; Rausch et al. 2008). And here it is important to note that there is a positive association observed between the survivor rating of severity or distress of events and reports of PTG (Alvaro and McFarland 1997; McMillen et al. 1997; Tedeschi 1999; Tedeschi and Calhoun 1996). In sum it is clear that this gender difference in trauma perception may cause a similar gender difference in PTG experience (Jin et al. 2014).

Rumination

Deliberate rumination, which is productive and contemplative in nature on productive issues on personal strength, and importance of social connectedness associated to traumatic events are playing a greater role in PTG (Calhoun and Tedeschi 2004; Janoff-Bulman et al. 2006; Vishnevsky et al. 2010). Reflective rumination of the traumatic event predicted positive effect and well-being among survivors of breast cancer (Segerstrom et al. 2003). Treynor et al. (2003) evidenced the increased ruminating tendency of the women; it included both deliberate and reflective rumination. Meta-analytic study conducted by Vishnevsky et al. (2010) also supported this explanation and pointed out the possible mediating role of ruminative thought in the relationship between female gender and PTG.

Coping Behaviour

A self-reported study conducted among 100 survivors of an earthquake or volcano eruption in Indonesia showed that coping was significantly associated to PTG. Female participants expressed higher levels of PTG than male participants, and additional findings show that similar gender difference was observed in coping behaviour also—both avoidance coping and approach coping were significantly higher in female groups. However, approach coping was found to be significantly effective in female participants (Akbar and Witruk 2016), additional evidences come from the study conducted by Helgeson et al. (2006) and Butler et al. (2005), which points that PTG was related to emotion-focused coping strategies, acceptance and positive reframing. These possibilities are further strengthened by the results of the studies by De Ridder (2000), Thoits (1991) and (Vingerhoets and Van Heck 1990) that women are more likely to utilize emotion-focused coping.

Self-disclosure

High degree of self-disclosure was found to associate with lower levels of PTSD and lower stress in turn associated to faster recovery and this recovery mediated the relationship between disclosure and PTG among people with early psychosis (Pietruch and Jobson 2012). Crispin et al. (2012) reported that people disclose after a traumatic event scored higher in interpersonal relations dimension of PTG, this result was similar in both American and Japanese sample. A number of studies have reported that men engage in relatively less emotional self-disclosure than women (Dindia and Allen 1992; Purves and Erwin 2004). In summary, it is evidenced that an existing gender difference in self-disclosing nature following a traumatic incident may also be a factor in the observed gender difference in PTG.

New Social Roles

Wortman et al. (1993) reported a significant gender difference in the perceived personal growth among 800 widowers and widows who lost their spouse. Women reported that they felt they had become stronger person after the death of their spouse (Wortman 2004). As a result of the spouses' death, women are required to perform new tasks and responsibilities, which may enhance their feelings of competence and they may develop new coping skills and awareness of their own strengths (Carr 2002).

Socialization and Social Support

Seeking help from others and PTG are significantly associated (Akbar and Witruk 2016; Swickert and Hittner 2009; Vishnevsky et al. 2010). Socialization processes of male and female are different. Females are expected to get help from others to cope with the problems, whereas a similar behaviour from male is discouraged because it signifies weakness (Akbar and Witruk 2016). Pederson and Vogel (2007) suggest that women are more likely to seek professional help than men. Individuals who perceive higher levels of stress in an event in association with higher social support is found to experience PTG (Calhoun and Tedeschi 2008). Research shows a similar tendency that women tend to perceive the events more distressing and also tend to receive higher levels of social support during the difficult moments. Experience of adverse events in the presence of social support may result in reported feelings of closeness towards the significant others (McMillen 1999; Taylor et al. 2000). Being an adult daughter caregiver of parents with cancer, they did not enjoy the greater positive outcomes even amidst reported social support in a study conducted by Texeira and Maria (2014). This conflicting finding will be explained by Linley and Joseph (2004), and according to their study there might be a significant relationship between positive personal changes and satisfaction with social support, but not social support itself. This particular group is to experience frequent role changing from being a caregiver of sick parent, to wife and parent of their own family. Decreased social activity participation, increased family conflict and decreased couple intimacy can reduce the quality and satisfaction of the received social support.

Empathy

A multi-mediation model of PTG explaining the role of social support and empathy reported that empathy played a more important role than social support in the relationship between gender and PTG. Here empathy specifies one's ability to understand another's emotional state and to respond in similar ways of their response and study further adds that women are not necessarily possessing greater accuracy in empathic understanding, but they display greater empathy in responding to the people around them than the men do (Chakrabarti and Baron-Cohen 2006; Neff and Karney 2005). Additional evidences show that women self-report higher levels of empathy than men (Davis 1994) and girls develop a theory of mind, which is important in PTG, earlier than boys (Putallaz et al. 1995; Walker 2005). In summary, it is evident that greater levels of empathy also play a role in the gender difference of PTG.

Biological Factors

Social neuroscience studies reported that the hormone oxytocin, which is related to oestrogen, the female sex hormone, adds to empathic responses (Carter et al. 2008). Singer et al. (2004) show that anterior insula and anterior cingulate, part of pain matrix, are comparatively less active in men, when they view an individual in pain. Resulting in lesser empathic reaction in men, since empathy is related to PTG, these neurological factors may also contribute to the gender difference observed (Swickert et al. 2012).

Methodological Incompetence

Possibly the observed gender differences may be at times due to methodological incompetence and errors in the research process. Since no research is error free, it is important to consider the possible methodological flaws that would have contributed to gender difference in PTG. It does not mean that the given results are not true, but it mentions the fact that it can be the result of certain flaws too. Many reported studies were cross-sectional studies and most of them utilized the self-report measures of PTG for measuring it among various groups. Frazier and Kaler (2006) and Ransom et al. (2008) evidenced the possibility that the scores on self-report measures may not correspond to actual growth. Additional validation efforts from the part of researchers are also reported as it is given in Tomich and Helgeson (2004), Park et al. (2005) and Weiss (2002). But these sort of additional efforts are not evident in every study.

Gender Composition

Another difficulty observed with the studies was that of gender composition, many studies reported a large gap between the number of male and female participants. Polatinsky and Esprey's study (2000) on assessing gender difference in PTG of parents who lost the children was conducted among 48 females and 17 males. Barskova and Oesterreich (2009) and Helgeson et al. (2006) also opinionated similarly (Vishnevsky et al. 2010).

Publication Bias

Meta-analysis conducted by Vishnevsky et al. (2010) compared the effect sizes of gender difference in the published and unpublished studies and observed a slight increase in effect size in case of published studies; however this result was not significant due to the overlapping confidence intervals.

Implications

These findings point out the need of future research designed specifically to examine the possible factors associated to gender difference in PTG. Understanding the mechanisms leading to higher levels of PTG among women will be of bidirectional utility in nature because it may explore the additional factors and processes contributing to PTG and thereby increase the understanding of the PTG experiences and its utilities. This also points to the need of conducting future studies with strong methodology, which address the possible flaws of the reported literatures as well. These findings specify the need of future studies to develop and validate gender-specific interventions that facilitate PTG. It also points to the need of reflective listening in clinical setting. It reminds the clinicians about the required special effort when dealing with men experiencing traumatic stress.

Conclusion

Trauma can be natural or manmade. The consequences are dreary, and among those dreary consequences, lies the possibilities of growth and development. PTG is a comparatively new focus in the field of psychology. It is considered as a salutogenic consequence of trauma experienced. Trauma is any event which exceeds the coping ability of the individual and this sort of situation may shatter the otherwise strong beliefs of the human life and creates a more real and strong belief about the life. This particular process brings changes to the five major dimensions of one's life, namely attitude towards life, strong, warm interpersonal relationships, new possibilities, spirituality and strength perception. Another important result that requires attention is that of observed gender difference in the levels of PTG among men and women. Here the present study focuses on the available literature concerning the gender difference in PTG and tries to analyse the possible factors contributing to this result. It is important to understand the underlying mechanism of gender difference in PTG since it is instrumental in developing tailor-made clinical techniques to promote PTG in the clients. Also put forward the need of more empirical research studies to address the causes of gender difference in the PTG.

PTG is a salutogenic outcome of traumatic experience, and it is evidenced to enhance one's positive outcomes in life and well-being. Clinical utility of PTG justifies the proliferation in the number of scientific studies in this particular variable. Among the study findings, many studies equivocally reported a small but significant gender difference in PTG reports between men and women. Women reported higher growth. The present study examined many of the available literature and focused on the reported affective, cognitive, social, biological and methodological causes of this reported gender difference, and this study also points to the need of methodologically strong, unbiased and open-minded researches and gender-specific intervention.

References

Affleck, G. and Tennen, H. (1991). "The Effect of Newborn Intensive Care on Parents' Psychological Well-being", *Children's Health Care*, Vol. 20, No. 1, pp. 6–14.

Affleck, G., Tennen, H., Croog, S., and Levine, S. (1987). "Causal Attribution, Perceived Benefits, and Morbidity after a Heart Attack: An 8-Year Study", *Journal of Consulting and Clinical Psychology*, Vol. 55, No. 1, pp. 29–35.

Akbar, Z. and Witruk, E. (2016). "Coping Mediates the Relationship Between Gender and Posttraumatic Growth", *Procedia-Social and Behavioral Sciences*, Vol. 217, pp. 1036–1043.

Aldwin, C. M., Levenson, M. R., and Spiro, A. (1994). "Vulnerability and Resilience to Combat Exposure: Can Stress Have Lifelong Effects?", *Psychology and Aging*, Vol. 9, No. 1, pp. 34–44.

Alvaro, C. and McFarland, C. (1997). *Victims' self-improvement perceptions: Enhancing the present of derogating the past.* Paper presented at the poster presentation at the Annual Meeting of the American Psychological Association, Chicago.

Anderson, K. M. and Manuel, G. (1994). "Gender Differences in Reported Stress Response to the Loma Prieta Earthquake", *Sex Roles*, Vol. 30, Nos. 9/10, pp. 725–733.

Ariyabandu, M. M. and Wickramasinghe, M. (2005). *Gender Dimensions in Disaster Management: A Guide for South Asia*: New Delhi: Zubaan.

Arpawong, T. E., Sussman, S., Milam, J. E., Unger, J. B., Land, H., Sun, P. and Rohrbach, L. A. (2015). "Post-traumatic Growth, Stressful Life Events, and Relationships with Substance Use Behaviors among Alternative High School Students: A Prospective Study", *Psychology & Health*, Vol. 30, No. 4, pp. 475–494.

Barskova, T. and Oesterreich, R. (2009). "Post-traumatic Growth in People Living with a Serious Medical Condition and Its Relations to Physical and Mental Health: A Systematic Review", *Disability and Rehabilitation*, Vol. 31, No. 21, pp. 1709–1733.

Bellizzi, K. M. (2004). "Expressions of Generativity and Posttraumatic Growth in Adult Cancer Survivors", *The International Journal of Aging and Human Development*, Vol. 58, No. 4, pp. 267–287.

Bensimon, M. (2012). "Elaboration on the Association between Trauma, PTSD and Posttraumatic Growth: The Role of Trait Resilience", *Personality and Individual Differences*, Vol. 52, No. 7, pp. 782–787.

Bhat, R. M. and Rangaiah, B. (2015). "The Impact of Conflict Exposure and Social Support on Posttraumatic Growth among the Young Adults in Kashmir", *Cogent Psychology*, Vol. 2, No. 1, http://dx.doi.org/10.1080/23311908.2014.1000077

Blaikie, P., Cannon, T., Davis, I., and Wisner, B. (1994). *At Risk: Natural Hazards, People's Vulnerability and Disasters*. London: Routledge.

Bower, J. E., Kemeny, M. E., Taylor, S. E., and Fahey, J. L. (1998). "Cognitive Processing, Discovery of Meaning, CD4 Decline, and AIDS-Related Mortality among Bereaved HIV-Seropositive Men", *Journal of Consulting and Clinical Psychology*, Vol. 66, No. 6, pp. 979–986. http://dx.doi.org/10.1037/0022-006X.66.6.979.

Butler, L. D., Blasey, C. M., Garlan, R. W., McCaslin, S. E., Azarow, J., Chen, X.H., Desjardins, J.C., DiMiceli, S., Seagraves, D.A., Kramer, H.C., Spiegel, D., & Hastings, T. A. (2005). "Posttraumatic Growth Following the Terrorist Attacks of September 11, 2001: Cognitive, Coping, and Trauma Symptom Predictors in an Internet Convenience Sample", *Traumatology*, Vol. 11, No. 4, pp. 247–267. http://dx.doi.org/10.1177/153476560501100405

Cadell, S., Hemsworth, D., Smit Quosai, T., Steele, R., Davies, E., Liben, S. Straatman, L.,&Siden, H. (2014). "Posttraumatic Growth in Parents Caring for a Child with a Life-Limiting Illness: A Structural Equation Model", *American Journal of Orthopsychiatry*, Vol. 84, No. 2, pp. 123–133. http://dx.doi.org/10.1037/h0099384

Calhoun, L. G. and Tedeschi, R. G. (1998). "Posttraumatic Growth: Future Directions", in *Posttraumatic Growth: Positive Changes in the Aftermath of Crisis*, ed. R. G. Tedeschi, C. Park, and L. G. Calhoun. Mahwah, NJ: Lawrence Erlbaum Associates. pp. 215–238.

Calhoun, L. G. and Tedeschi, R. G. (2001). "Posttraumatic Growth", *Corsini Encyclopedia of Psychology*. Vol. 1, No.(2) doi: 10.1002/9780470479216.corpsy0698

Calhoun, L. G. and Tedeschi, R. G. (2004). Author's response: "The Foundations of Posttraumatic Growth: New Considerations", *Psychological inquiry*, Vol. 15, No. 1, pp. 93–102.

Calhoun, L. G. and Tedeschi, R. G. (2008). "The Paradox of Struggling with Trauma: Guidelines for Practice and Directions for Research", in *Trauma, Recovery, and Growth: Positive Psychological Perspectives on Posttraumatic Stress*, ed. S. Joseph and P. A. Linley. Hoboken, NJ: Wiley. pp. 325–337.

Carr, D. (2002). *Life after Death: Gender Differences in Resilience among the Elderly Widowed.* Paper presented at the 55th Annual Scientific Meeting, Gerontological Society of America, Boston, MA.

Carter, C. S., Grippo, A. J., Pournajafi-Nazarloo, H., Ruscio, M. G., and Porges, S. W. (2008). "Oxytocin, Vasopressin and Sociality", *Progress in Brain Research*, Vol. 170, pp. 331–336.

Chakrabarti, B. and Baron-Cohen, S. (2006). "Empathizing: Neurocognitive Developmental Mechanisms and Individual Differences", *Progress in Brain Research*, Vol. 156, pp. 403–417.

Crispin, C., Sawa, M., and Taku, K. (2012). "The Effect of Disclosure on Relating to Others Domain of Posttraumatic Growth". Retrieved from https://kanakotaku.files.wordpress.com/2012/08/crispin_sawa_taku_m-o-m-paper_20121.pdf

Curbow, B., Somerfield, M. R., Baker, F., Wingard, J. R. and Legro, M. W. (1993). "Personal Changes, Dispositional Optimism, and Psychological Adjustment to Bone Marrow Transplantation", *Journal of Behavioral Medicine*, Vol. 16, No. 5, pp. 423–443.

Davis, M. H. (1994). *Empathy: A Social Psychological Approach*. Boulder, CO: Westview Press.

De Ridder, D. (2000). "Gender, Stress and Coping: Do Women Handle Stressful Situations Differently from Men", *Women, Health and the Mind*, pp. 115–135.

Decker, L. R. (1993). "The Role of Trauma in Spiritual Development", *Journal of Humanistic Psychology*, Vol. 33, No. 4, pp. 33–46.

Dindia, K. and Allen, M. (1992). "Sex Differences in Self-disclosure: A Meta-analysis", *Psychological Bulletin*, Vol. 112, No. 1, pp. 106–124. http://dx.doi.org/10.1037/0033-2909.112.1.106

Dunigan, J. T., Carr, B. I., and Steel, J. L. (2007). "Posttraumatic Growth, Immunity and Survival in Patients with Hepatoma", *Digestive Diseases and Sciences*, Vol. 52, No. 9, pp. 2452–2459.

Edmonds, S. and Hooker, K. (1992). "Perceived Changes in Life Meaning Following Bereavement", *OMEGA: Journal of Death and Dying*, Vol. 25, No. 4, pp. 307–318.

Felten-Biermann, C. (2006). "Gender and Natural Disaster: Sexualized Violence and the Tsunami", *Development*, Vol. 49, No. 3, pp. 82–86.

Frazier, P. A., Conlon, A., and Glaser, T. (2001). "Positive and Negative Life Changes Following Sexual Assault", *Journal of Consulting and Clinical Psychology*, Vol. 69, No. 6, pp. 1048–1055. http://dx.doi.org/10.1037/0022-006X.69.6.1048

Frazier, P. A. and Kaler, M. E. (2006). "Assessing the Validity of Self-reported Stress-related Growth", *Journal of Consulting and Clinical Psychology*, Vol. 74, No. 5, pp. 859-869. http://dx.doi.org/10.1037/0022-006X.74.5.859

Gerrish, N., Dyck, M. J., and Marsh, A. (2009). "Post-traumatic Growth and Bereavement", *Mortality*, Vol. 14, No. 3, pp. 226–244.

Halcomb, E., Daly, J., Davidson, P., Elliott, D., and Griffiths, R. (2005). "Life Beyond Severe Traumatic Injury: An Integrative Review of the Literature", *Australian Critical Care*, Vol. 18, No. 1, pp. 17–24.

Helgeson, V. S., Reynolds, K. A., and Tomich, P. L. (2006). "A Meta-analytic Review of Benefit Finding and Growth", *Journal of Consulting and Clinical Psychology*, Vol. 74, No. 5, pp. 797–816. http://dx.doi.org/10.1037/0022-006X.74.5.797

Ho, S. M., Chan, C. L., and Ho, R. T. (2004). "Posttraumatic Growth in Chinese Cancer Survivors", *Psycho-Oncology*, Vol. 13, No. 6, pp. 377–389.

Hooper, L. M. (2003). "Parentification, Resiliency, Secure Adult Attachment Style, and Differentiation of Self as Predictors of Growth Among College Students", *Dissertations Abstracts International*, Vol. 64, p. 1493. Retrieved from: https://elibrary.ru/item.asp?id=5712014 on 12/08/16.

Jang, L.-J. (2005). *The 921 Earthquake: A Study of the Effects of Taiwanese Cultural Factors on Resilience*. Doctoral Dissertations, University of Denver.

Janoff-Bulman, R. (2010). Shattered Assumptions: Towards a new Psychology of Trauma. New York: Simon and Schuster.

Janoff-Bulman, R., Calhoun, L. G., and Tedeschi, R. G. (2006). "Schema-Change Perspectives on Posttraumatic Growth", *Handbook of Posttraumatic Growth: Research and Practice*. Mahwah, NJ: Lawrence Erlbaum Associates. pp. 81–99.

Jin, Y., Xu, J., and Liu, D. (2014). "The Relationship between Posttraumatic Stress Disorder and Posttraumatic Growth: Gender Differences in PTG and PTSD Subgroups", *Social Psychiatry and Psychiatric Epidemiology*, Vol. 49, No. 12, pp. 1903–1910.

Joseph, S. (2009). "Growth Following Adversity: Positive Psychological Perspectives on Posttraumatic Stress". *Psihologijske teme*, Vol. 18, No. 2, pp. 335–344.

Joseph, S. and Linley, P. A. (2005). "Positive Adjustment to Threatening Events: An Organismic Valuing Theory of Growth through Adversity", *Review of General Psychology*, Vol. 9, No. 3, pp. 262–280. DOI: 10.1037/1089-2680.9.3.262

Joseph, S., Linley, P. A., Andrews, L., Harris, G., Howle, B., Woodward, C., and Shevlin, M. (2005). "Assessing Positive and Negative Changes in the Aftermath of Adversity: Psychometric Evaluation of the Changes in Outlook Questionnaire", *Psychological Assessment*, Vol. 17, No. 1, pp. 70–80. doi:http://dx.doi.org/10.1037/1040-3590.17.1.70

Joseph, S., Murphy, D., and Regel, S. (2012). "An Affective–Cognitive Processing Model of Post-traumatic Growth", *Clinical Psychology & Psychotherapy*, Vol. 19, No. 4, pp. 316–325.

Keane, T. M., Marshall, A. D., and Taft, C. T. (2006). "Posttraumatic Stress Disorder: Etiology, Epidemiology, and Treatment Outcome", *Annual Review of Clinical Psychology*, Vol. 2, pp. 161–197.

Kelly, J. (2004). "Spirituality as a Coping Mechanism", *Dimensions of Critical Care Nursing*, Vol. 23, No. 4, pp. 162–168.

Lehman, D. R., Davis, C. G., DeLongis, A., Wortman, C. B., Bluck, S., Mandel, D. R., and Ellard, J. H. (1993). "Positive and Negative Life Changes Following Bereavement and Their Relations to Adjustment", *Journal of Social and Clinical Psychology*, Vol. 12, No. 1, pp. 90–112.

Lindstrom, C. M., Cann, A., Calhoun, L. G., and Tedeschi, R. G. (2013). "The Relationship of Core Belief Challenge, Rumination, Disclosure, and Sociocultural Elements to Posttraumatic Growth". *Psychological Trauma: Theory, Research, Practice, and Policy*, Vol. 5, No. 1, pp. 50–55. doi: 10.1037/a0022030.

Linley, P. A. and Joseph, S. (2004). "Positive Change Following Trauma and Adversity: A Review", *Journal of Traumatic Stress*, Vol. 17, No. 1, pp. 11–21.

Maercker, A. and Langner, R. (2001). "Posttraumatic Personal Growth: Validation of German Versions of Two Questionnaires", *Diagnostica*, Vol. 47, No. 3, pp. 153–162.

Maercker, A. and Zoellner, T. (2004). "The Janus Face of Self-perceived Growth: Toward a Two-Component Model of Posttraumatic Growth", *Psychological Inquiry*, Vol. 15, No. 1, pp. 41–48.

Malhotra, M. and Chebiyan, S. (2016). "Posttraumatic Growth: Positive Changes Following Adversity: An Overview", *International Journal of Psychology and Behavioral Sciences*, Vol. 6, No. 3, pp. 109–118.

McIntosh, D. N., Silver, R. C., and Wortman, C. B. (1993). "Religion's Role in Adjustment to a Negative Life Event: Coping with the Loss of a Child", *Journal of Personality and Social Psychology*, Vol. 65, No. 4, pp. 812–821.

McMillen, J. C. (1999). "Better for It: How People Benefit from Adversity", *Social Work*, Vol. 44, No. 5, pp. 455–468.

McMillen, J. C. and Fisher, R. H. (1998). "The Perceived Benefit Scales: Measuring Perceived Positive Life Changes after Negative Events", *Social Work Research*, Vol. 22, No. 3, pp. 173–187.

McMillen, J. C., Smith, E. M., and Fisher, R. H. (1997). "Perceived Benefit and Mental Health after Three Types of Disaster", *Journal of Consulting and Clinical Psychology*, Vol. 65, No. 5, pp. 733–739. doi: http://dx.doi.org/10.1037/0022-006X.65.5.733.

McMillen, J. C., Zuravin, S., and Rideout, G. (1995). "Perceived Benefit from Child Sexual Abuse", *Journal of Consulting and Clinical Psychology*, Vol.

63, No. 6, pp. 1037–1043. http://dx.doi.org/10.1037/0022-006X.63.6.1037

Milam, J., Ritt-Olson, A., Tan, S., Unger, J., and Nezami, E. (2005). "The September 11th 2001 Terrorist Attacks and Reports of Posttraumatic Growth among a Multi-Ethnic Sample of Adolescents", *Traumatology*, Vol. 11, No. 4, pp. 233–246. http://dx.doi.org/10.1177/153476560501100404

Miles, M. S. and Crandall, E. K. B. (1983). "The Search for Meaning and Its Potential for Affecting Growth in Bereaved Parents", in *Coping with Life Crises*. Springer. pp. 235–243.

Mooney, M. F., Johal, S., Paton, D., Tarrant, R., and Johnston, D. (2012). "Surviving and Thriving: An Introduction to Childhood and Youth Post-Disaster Recovery in the Context of the Canterbury Earthquakes of 2010–2012". School of Psychology, Massey University. http://mro.massey.ac.nz/handle/10179/2645.

Morris, B. A. and Shakespeare-Finch, J. (2011). "Rumination, Post-traumatic Growth, and Distress: Structural Equation Modelling with Cancer Survivors", *Psycho-Oncology*, Vol. 20, No. 11, pp. 1176–1183.

Neff, L. A. and Karney, B. R. (2005). "Gender Differences in Social Support: A Question of Skill or Responsiveness? *Journal of Personality and Social Psychology*, Vol. 88, No. 1, pp. 79–90. http://dx.doi.org/10.1037/0022-3514.88.1.79

Olff, M., Langeland, W., Draijer, N., and Gersons, B. P. (2007). "Gender Differences in Posttraumatic Stress Disorder", *Psychological Bulletin*, Vol. 133, No. 2, pp. 183–204. http://dx.doi.org/10.1037/0033-2909.133.2.183

Park, C. L., Cohen, L. H., and Murch, R. L. (1996). "Assessment and Prediction of Stress-Related Growth", *Journal of Personality*, Vol. 64, No. 1, pp. 71–105.

Park, C. L., Mills-Baxter, M. A., and Fenster, J. R. (2005). "Post-Traumatic Growth from Life's Most Traumatic Event: Influences on Elders' Current Coping and Adjustment", *Traumatology*, Vol. 11, No. 4, pp. 297–306. http://dx.doi.org/10.1177/153476560501100408

Patrick, J. H. and Henrie, J. (2016). "Up From the Ashes: Age and Gender Effects on Post-Traumatic Growth in Bereavement", *Women & Therapy*, Vol. 39, Nos. 3/4, pp. 296–314.

Pederson, E. L. and Vogel, D. L. (2007). "Male Gender Role Conflict and Willingness to Seek Counseling: Testing a Mediation Model on College-aged Men", *Journal of Counseling Psychology*, Vol. 54, No. 4, pp. 373-384. http://dx.doi.org/10.1037/0022-0167.54.4.373

Peres, J. F., Moreira-Almeida, A., Nasello, A. G., and Koenig, H. G. (2007). "Spirituality and Resilience in Trauma Victims", *Journal of Religion and Health*, Vol. 46, No. 3, pp. 343–350.

Phipps, S., Long, A. M., and Ogden, J. (2007). "Benefit Finding Scale for Children: Preliminary Findings from a Childhood Cancer Population", *Journal of Pediatric Psychology*, Vol. 32, No. 10, pp. 1264–1271.

Pietruch, M. and Jobson, L. (2012). "Posttraumatic Growth and Recovery in People with First Episode Psychosis: An Investigation into the Role of Self-disclosure", *Psychosis*, Vol. 4, No. 3, pp. 213–223.

Polatinsky, S. and Esprey, Y. (2000). "An Assessment of Gender Differences in the Perception of Benefit Resulting from the Loss of a Child", *Journal of Traumatic Stress*, Vol. 13, No. 4, pp. 709–718.

Prati, G. and Pietrantoni, L. (2009). "Optimism, Social Support, and Coping Strategies as Factors Contributing to Posttraumatic Growth: A Meta-analysis", *Journal of Loss and Trauma*, Vol. 14, No. 5, pp. 364–388.

Purves, D. G. and Erwin, P. G. (2004). "Post-traumatic Stress and Self-disclosure", *The Journal of Psychology*, Vol. 138, No. 1, pp. 23–34.

Putallaz, M., Hellstern, L., Sheppard, B. H., Grimes, C. L., and Glodis, K. A. (1995). "Conflict, Social Competence, and Gender: Maternal and Peer Contexts", *Early Education and Development*, Vol. 6, No. 4, pp. 433–447.

Quarantelli, E. L. (1994). "Looting and Antisocial Behavior in Disasters". Encyclopedia of Criminology. Disaster Research Centre. http://udspace.udel.edu/handle/19716/590

Ramos, C. and Leal, I. (2013). "Posttraumatic Growth in the Aftermath of Trauma: A Literature Review about Related Factors and Application Contexts", *Psychology, Community & Health*, Vol. 2, No. 1, pp. 43–54.

Ransom, S., Sheldon, K. M., and Jacobsen, P. B. (2008). "Actual Change and Inaccurate Recall Contribute to Posttraumatic Growth Following Radiotherapy", *Journal of Consulting and Clinical Psychology*, Vol. 76, No. 5, pp. 811–819. http://dx.doi.org/10.1037/a0013270.

Rausch, S. M., Auerbach, S. M., and Gramling, S. E. (2008). "Gender and Ethnic Differences in Stress Reduction, Reactivity, and Recovery", Sex Roles, Vol. 59, No. 9, 726–737. doi 10.1007/s11199-008-9460-0

Rees, S., Pittaway, E., and Bartolomei, L. (2005). "Waves of Violence: Women in Post-tsunami Sri Lanka", *Australasian Journal of Disaster and Trauma Studies*, pp. 1–6.

Rogers, C. R. (1951). "Perceptual Reorganization in Client-centered Therapy", in Perception: An Approach to Personality, ed. R. R. Blake and G. V. Ramsey. New York: Ronald Press. pp. 307–327.

Schaefer, A. J. and Moos, R. H. (1992). "Life Crises and Personal Growth", in *Personal Coping Theory, Research, and Application*, ed. B. N. Carpenter. Westport: Praeger. pp. 149–170.

Scrignaro, M., Barni, S., and Magrin, M. E. (2011). "The Combined Contribution of Social Support and Coping Strategies in Predicting Post-traumatic Growth: A Longitudinal Study on Cancer Patients", *Psycho-Oncology*, Vol. 20, No. 8, pp. 823–831.

Segerstrom, S. C., Stanton, A. L., Alden, L. E., and Shortridge, B. E. (2003). "A Multidimensional Structure for Repetitive Thought: What's on Your Mind, and How, and How Much?", *Journal of Personality and Social Psychology*, Vol. 85, No. 5, pp. 909–921.

Sheldon, K. M., Arndt, J., and Houser-Marko, L. (2003). "In Search of the Organismic Valuing Process: The Human Tendency to Move towards Beneficial Goal Choices", *Journal of Personality*, Vol. 71, No. 5, pp. 835–869.

Singer, T., Seymour, B., O'Doherty, J., Kaube, H., Dolan, R. J., and Frith, C. D. (2004). "Empathy for Pain Involves the Affective but Not Sensory Components of Pain", *Science*, Vol. 303, No. 5661, pp. 1157–1162.

Solomon, Z. and Dekel, R. (2007). "Posttraumatic Stress Disorder and Posttraumatic Growth among Israeli ex-POWs", *Journal of Traumatic Stress*, Vol. 20, No. 3, pp. 303–312.

Swickert, R. and Hittner, J. (2009). "Social Support Coping Mediates the Relationship between Gender and Posttraumatic Growth", *Journal of Health Psychology*, Vol. 14, No. 3, pp. 387–393.

Swickert, R., Hittner, J. B., and Foster, A. (2012). "A Proposed Mediated Path between Gender and Posttraumatic Growth: The Roles of Empathy and Social Support in a Mixed-Age Sample", *Psychology*, Vol. 3, No. 12, pp. 1142–1147. http://dx.doi.org/10.4236/psych.2012.312A168.

Taku, K., Cann, A., Calhoun, L. G., and Tedeschi, R. G. (2008). "The Factor Structure of the Posttraumatic Growth Inventory: A Comparison of Five Models Using Confirmatory Factor Analysis", *Journal of Traumatic Stress*, Vol. 21, No. 2, pp. 158–164.

Taku, K., Cann, A., Tedeschi, R. G., and Calhoun, L. G. (2009). "Intrusive Versus Deliberate Rumination in Posttraumatic Growth across US and Japanese Samples", *Anxiety, Stress, & Coping*, Vol. 22, No. 2, pp. 129–136.

Talbot, K. (1999). "Mothers now Childless: Personal Transformation after the Death of an Only Child", *OMEGA: Journal of Death and Dying*, Vol. 38, No. 3, pp. 167–186.

Taylor, S. E. (1983). "Adjustment to Threatening Events: A Theory of Cognitive Adaptation", *American Psychologist*, Vol. 38, No. 11, pp. 1161–1173.

Taylor, S. E., Kemeny, M. E., Reed, G. M., Bower, J. E., and Gruenewald, T. L. (2000). "Psychological Resources, Positive Illusions, and Health". *Advances In Mind – Body Medicine*, Vol. 17, No. 1 pp. 48–50. https://elibrary.ru/item.asp?id=4113614.

Tedeschi, R. G. (1999). "Violence Transformed: Posttraumatic Growth in Survivors and Their Societies", *Aggression and Violent Behavior*, Vol. 4, No. 3, pp. 319–341.

Tedeschi, R. G. and Calhoun, L. G. (1995). *Trauma and Transformation: Growing in the Aftermath of Suffering*. New York: Sage Publications.

Tedeschi, R. G. and Calhoun, L. G. (1996). "The Posttraumatic Growth Inventory: Measuring the Positive Legacy of Trauma", *Journal of Traumatic Stress,* Vol. 9, No. 3, pp. 455–471.

Tedeschi, R. G., & Calhoun, L. G. (2004). Posttraumatic growth: Conceptual foundations and empirical evidence. *Psychological Inquiry, 15*(1), 1-18. doi:10.1207/s15327965pli1501_01

Tempelaar, R., de Haes, J. C., De Ruiter, J., Bakker, D., Van den Heuvel, W., and Van Nieuwenhuijzen, M. (1989). "The Social Experiences of Cancer Patients under Treatment: A Comparative Study", *Social Science & Medicine*, Vol. 29, No. 5, pp. 635–642.

Texeira, R. J. and Maria, G. P. (2014). "Family Caregiver Support to Cancer Patients: The Relevance of Adult Children", in *The Praeger Handbook on Women's Cancers: Personal and Psychosocial Insights*, ed. M. A. Paludi. Santa Barbara, CA: Praeger. pp. 305–333.

Thoits, P. A. (1991). "Gender Differences in Coping with Emotional Distress", in *The Social Context of Coping*. Springer. pp. 107–138.

Thompson, S. C. (1991). "The Search for Meaning Following a Stroke", *Basic and Applied Social Psychology*, Vol. 12, No. 1, pp. 81–96.

Tolin, D. F. and Foa, E. B. (2006). "Sex Differences in Trauma and Posttraumatic Stress Disorder: A Quantitative Review of 25 Years of research", Psychological Bulletin, Vol. 132, No. 6, pp. 959–992. doi: 10.1037/0033-2909.132.6.959

Tomich, P. L. and Helgeson, V. S. (2004). "Is Finding Something Good in the Bad Always Good? Benefit Finding among Women with Breast Cancer", Health Psychology, Vol. 23, No. 1, pp. 16–23. doi:http://dx.doi.org/10.1037/0278-6133.23.1.16

Tomm, K. (1987). "Interventive Interviewing: Part II. Reflexive Questioning as a Means to Enable Self-healing", *Family Process*, Vol. 26, No. 2, pp. 167–183.

Treynor, W., Gonzalez, R., and Nolen-Hoeksema, S. (2003). "Rumination Reconsidered: A Psychometric Analysis", *Cognitive Therapy and Research*, Vol. 27, No. 3, pp. 247–259.

Vingerhoets, A. J. and Van Heck, G. L. (1990). "Gender, Coping and Psychosomatic Symptoms", *Psychological Medicine*, Vol. 20, No. 1, pp. 125–135.

Vishnevsky, T., Cann, A., Calhoun, L. G., Tedeschi, R. G., and Demakis, G. J. (2010). "Gender Differences in Self-reported Posttraumatic Growth: A Meta-analysis", *Psychology of Women Quarterly*, Vol. 34, No. 1, pp. 110–120.

Walker, S. (2005). "Gender Differences in the Relationship between Young Children's Peer-related Social Competence and Individual Differences in Theory of Mind", *The Journal of Genetic Psychology*, Vol. 166, No. 3, pp. 297–312.

Weiss, H. M. (2002). "Introductory Comments: Antecedents of Emotional Experiences at Work", *Motivation and Emotion*, Vol. 26, No. 1, pp. 1–2.

Widows, M. R., Jacobsen, P. B., Booth-Jones, M., and Fields, K. K. (2005). "Predictors of Posttraumatic Growth Following Bone Marrow Transplantation for Cancer", *Health Psychology*, Vol. 24, No. 3, p. 266- 273.

Wortman, C. B. (2004). "Posttraumatic Growth: Progress and Problems", *Psychological Inquiry*, Vol. 15, No. 1, pp. 81–90.

Wortman, C. B., Silver, R. C., and Kessler, R. C. (1993). "The Meaning of Loss and Adjustment to Bereavement", in *Handbook of Bereavement: Theory, Research and Intervention*, ed. M. S. Stroebe, W. Stroebe, and R. O. Hansson. New York: Cambridge University Press. pp. 349–366.

Yalom, I. D. and Lieberman, M. A. (1991). "Bereavement and Heightened Existential Awareness", *Psychiatry*, Vol. 54, No. 4, pp. 334–345.

Zoellner, T. and Maercker, A. (2006). "Posttraumatic Growth in Clinical Psychology—A Critical Review and Introduction of a Two Component Model", *Clinical Psychology Review*, Vol. 26, No. 5, pp. 626–653.

PART IV

Assessing Preparedness, Recovery and Mitigation

10

Gender Sensitivity in Disaster Risk Management

Lesson Learnt from the Chennai Floods of 2015

Suriya S.

Introduction

Chennai metropolitan has survived the devastating floods during November to December 2015 and had drawn international attention. Merely calling it as a natural disaster is very limited as it exposed the poor potential of the city in its scope to face such hazards. Slum dwellers had been the primary victims of the flood and in the case of women, it has been a double burden for them to survive the natural disaster with both the class and gender card. Notwithstanding state's effort to regulate informal settlements, it is been critiqued that most of the resettlements are situated in areas that have high risks of floods (Coelho 2016). It became evident when the residents near the river and other low-lying areas were placed in vulnerable position. With the loss of income, they had struggled to access the basic necessities of food, clean water, shelter and sanitation. As the childcare relies on women by default and they had to deal with both physical and psychological stress and rose to the situation. However, it has been a wake-up call for us to be gender inclusive in disaster management as the similar pattern of rain is predicted in the coming years. The government has been taking precautions to handle such situation in the future and for it to be gender inclusive is the primary call of this study.

With the above-said notions, this study elucidates how gender intersects with other social and economic factors and influences disaster vulnerability, response and recovery. The importance of gender sensitivity in disaster risk management with reference to the crisis faced by slum dwellers and the initiatives taken by the government regarding the resettlement of slums forms the crux of this study.

Gender Mainstreaming in Disaster Risk Management

Gender turned out to be a social as well as a political category with a power imbalance between the sexes. It is the men who have edge over the power relations and play a major role as policy makers and as the right bearers. Contrarily, women's marginal participation in the mainstream governance restricted them as beneficiaries at the receiving ends (Fraser 1987). The reason behind this is the gendered socialization, where men and women have different life experience and consequently have different needs, concerns and priorities. Hence, the conventional attitude creates a difference between the two sexes in terms of capability, skills and power relations. The segregation influences occupation, governance and leadership and place women in a relatively vulnerable situation. These scenarios tag along times of 'disaster' and the women are at the receiving ends in terms of recovery.

According to Ross and Lott (2003) a natural disaster is a major adverse event resulting from natural processes of the earth; examples include floods, hurricanes, tornadoes, volcanic eruptions, earthquakes, tsunamis and other geologic processes. It can cause loss of life and property damage. Disaster risk management is the strategic approach to identify, assess and reduce the negative impacts of a natural disaster. It aims to reduce physical as well as socio-economic vulnerabilities caused by the natural hazards. Gender card which intersects with all three kinds of vulnerabilities required a special attention from the government as well as non-profit agencies. Therefore, the study of gender and vulnerability has to go combined with the disaster risk management. Gender concerns gradually drew attention in the global agenda of disaster risk management after the consistent global advocacy and awareness raised by international bodies headed by United Nations. The United Nations General

Assembly declared 1990–1999, the International Decade for Natural Disaster Reduction (IDNDR). Following that, the United Nations Office for Disaster Risk Reduction (UNISDR) was started in 1999. Since then, conferences and awareness programmes were conducted and the issues faced by women got identified and addressed. However, Convention for the Elimination of All Forms of Discrimination against Women (CEDAW) included the importance of having gendered perspective in managing natural hazards. It is a treaty held by United Nations which serves as the international blueprint to enhance gender equality. Article 7, General Recommendation of CEDAW, claims gender balance in planning, training and implementation efforts in environmental management and in natural disasters, in order to ensure non-discrimination of women and boost their empowerment (UN/ISDR 2001).

Having age-old membership in United Nations since 1942, India has been part of all specialized agencies of United Nations. It has strongly supported the policies and principles of United Nations. In the case of disaster risk management, India had been enacting laws and implementing agencies. It has enacted The Disaster Management Act 2005 to provide 'the effective management of disasters and for matters connected therewith or incidental thereto'. Accordingly, NDMA was established under the chairmanship of Prime Minister. National Executive Committee (NEC), State Disaster Management Authority (SDMA), District Disaster Management Authority (DDMA) and National Disaster Respond Force (NDRF) serve as executive agencies. National Institute of Disaster Management manages financial support for disaster mitigations at various levels. The importance of 'incorporating gender-sensitive and equitable approach to capacity development covering all aspects of Disaster risk management in the state districts and local level' has been mentioned in the all the plans (Government of India 2005).

Conversely, Tamil Nadu was one among the states to show poor potential towards disaster management. Following the public interest litigation regarding inefficiency of the state in the implementation of The Disaster Management Act 2005, the Supreme Court issued notice to Tamil Nadu and other states which witnessed natural calamities in the past (*The Times of India* 2013). Tamil Nadu has come across few hazards so far: tsunami in 2005; flood in 2015 and Vardah cyclone in 2016. All these hazards were devastating and brought huge loss to lives and property. This chapter focuses on Chennai floods 2015 and traces the gendered aspect of disaster risk carried out. It elucidates the flaws and finally posts the recommendation that needs to be taken into account in order to reduce women's vulnerability. The study adopts qualitative and participatory research as the methodology. Secondary sources including government reports, documents, guidelines issued by international bodies and newspaper articles were referred and in order to substantiate the sources and few in-depth interviews were conducted among women who reside in flood-affected slum of the city. The NGOs, community-based organizations, volunteers and officials gave their inputs to this study.

The Chennai Floods of 2015

Under the influence of depression in the Bay of Bengal, back to back spell of heavy rain wrecked Chennai during November to December 2015. It received 1,049 mm (41.3 in) of rainfall. The coastal area of the city, suburbs and nearby Union Territory had a record breaking rainfall since 1918 when 1,088 mm (42.8) was recorded. Particularly heavy downpour on 1 December forced the state to discharge water at the rate of 20,000 cubic feet per second to the Adyar River. The flood warning was issued by the officials, and the people who reside along the river were asked to move to the safe places. Kattupakkam near Chengalpattu received around 22 cm of rain, and the lakes in the region were overflowing and blocked the roads of Grand Southern Trunk between Chengalpattu and Chennai for the first time in 40 years. Trains and buses were suspended in the affected areas. Flights were cancelled as the airport was one among the most affected area. The city was marooned with water overflowing from rivers and reservoirs, destructed apartments, shopping malls, markets, office and educational institutions. Two hundred and eighty people were killed in the disaster. In one of the most shocking incidents, 18 patients in the intensive care unit at MIOT Hospital died due to the flood wrecking the generator which runs the life support devices. The victims reported that the water was released from the overflowing lakes without any warning (Kolappan 2015). People were stuck in the upper floors with no food, water and electricity for nearly two weeks. Food packs were airdropped by the army. Piles of garbage and sewage entered into houses via floods. Carcasses, snakes and rodents were spotted. People suffered falls, bites, allergies and skin aliments, and electrocution (Ramya 2015).

Government Rescue Operation

Forty army teams were involved in the rescue operation. A total of 1,200 NDRF personnel, 6 navy teams, 112 boats and 200 divers were involved in this operation. The NDRF proceeded with 131 inflatable and other kinds of boats and buoys and life jackets with the rescue and relief team. O. P. Singh was the Director General of the force (Relief web 2015). A meeting of the crisis management group (CMG) was held under the chairmanship of Union Home Secretary Rajiv Mehrishi to review the preparedness of the agency along with officials of Ministries of Defence, Food, Railways Agriculture, Health, Department of Telecommunication, National Disaster Management Authority NDMA, India Meteorological Department (IMD) and NDRF. Overall, 1.8 million people have been displaced. A total of 325,000 people were evacuated from the flood-battered areas. On the other hand, non-profit organizations, political parties, religious organizations, community-based associations and volunteers rendered their help in the circumstance. The state government announced free repair and service for two wheelers and three wheelers (Government of Tamil Nadu 2015). Central government announced ₹ 1,940 crore for Tamil Nadu. Civic infrastructure costing around ₹ 2,000 crores was estimated to be damaged by floods. Water receded and power was restored within the month.

Chennai Floods and Slums

Urban poor resided at the edge of rivers, railway tracks and in low-lying areas faced the worst of devastation. Tamil Nadu Slum Clearance Board has conducted the survey on slums. According to the survey, there was 51.85 per cent of an increase in slums from 2001 to 2014, that is, 306 enclaves in 1956 to 2,173 in 2014. This survey was conducted for Rajiv Awas Yojna announced by Government of India. The objective of this scheme is to establish a 'Slum Free India' by adopting a holistic approach to solve the problems of urban poor[1]. The objectives of this plan are to prevent the growth of slums and encroachments; to prevent the eviction of slum dwellers by private owners and to provide the slum families with the security of tenure and to provide basic amenities such as water supply, street lights, storm water drains and sewer line in the concerned areas. Accordingly, 7,500 crores were allocated to Tamil Nadu for building 1 lakh houses within the period of 2014–2022. Based on the survey, 2,173 slums were identified and around 304,980 households were located in the slums.

North Chennai has higher urban poor population. People migrate to avail employment opportunities there. Moreover, the availability of affordable houses attracts people in certain areas. The study reveals that half of the population belongs to SC/ST communities and 73 per cent of them are casual labourers. Four per cent of the households don't have toilet facilities and the people had to go for open defecation.

As a part of Rajiv Awas Yojna, entrepreneurial development programme was designed for women. It involves skill development training, credit linkage and marketing support with the help of non-governmental and community-based organizations. Financial support was given for sewing machine, iron box, wet grinder, push carts, mobile bunk, tricycle and toolkits for electricians, plumbers and beauticians. It also provides a fund for the *murukku* machine, ice box and umbrella to women in order to enhance their livelihood. Through self-help groups (SHG) women are encouraged to participate in Tsunami Emergency Assistance Projects such as luggage carrier, autos, solid waste management, departmental store and sanitary napkin production were implemented in Kannagi Nagar and Semmancheri where the slum dwellers were resettled (Draft Report 2014). Notwithstanding the fancy projects to ease slum dwellers, the government shifted people from the low-lying area and allotted housed in another low-lying area. Both the slum areas and the resettlements were washed away and people were left homeless. It caused many to get the sense of downward mobility in terms of their financial status. Most of them couldn't get rescued immediately and in certain areas, it took almost three to four day for the force to reach them. The rescued people were made to stay in nearby schools and colleges. Women, with the loss of income, had struggled to access the basic necessities of food, clean water, shelter and sanitation. Moreover, taking care of the children comes under women's duty and they had to deal with both physical and psychological stress to recover from the hazards.

As the majority of urban poor are casual labours, women had to carry out the lowest paid activities, mostly as domestic workers. They suffer the burden of balancing their household work as well. Media and alcohol and drugs influence the vulnerability of women and increase the domestic violence, workplace harassment, psychological stress, poor health and poverty. Studies reveal that there has been inadequate development in terms of social capital of economically active women in slum areas and they remain low skilled. They have very limited financial assets which continue to drain, and their participation in the formal economic is restricted to the consumers rather than producers (CARE 2006). Hence, in the case of natural hazards or riots or any other social kayos, women face a different set of problems.

Study Area

This study focused on the slums situated near Ambedkar Bridge, Mylapore, Chennai. It belongs to Ward Nos. 124 and 125. Total population comes around 5,000, and each street consists of average 245 families. Among those Meenambalpuram alone has 467 families. Egambaram has 20 housing boards. Other streets are Lutharpuram with 205 families; Munusamy Street, 123 families; Nattan Garden with about 203 families; PNK Garden with 100 families; Pallakilumanagar with 97 families; Selattaipuram with 134 families; Shanmugam Street with about 500 families; Nainar Road 150 families; Ganesapuram 610 families and Arunthathiyar Colony with around 150 families. One can identify the area with City Centre Multiplex Mall which is located nearby. It is one of the densely populated areas in Mylapore. It is a politically active place with meetings, rallies and riots happening often. It witnessed petty violence regularly. So a police booth has been set up at the entrance. It is one among the places where contractual agencies of Chennai Corporation dump the garbage collected in and around Mylapore and later it is taken to Perungudi.

Case Studies

The flood caused them to feel a sense of fall in the financial stability, and some people lost middle-class status. Hundreds of families in those slums lost their homes, a clear signifier of status. All of a sudden they were deprived of a place of comfort, the home they owned, their few clothes, furniture and vehicle that got carried away. Taking this scenario, the primary goal of having these case studies is to capture the consequences faced by women during natural hazards; what kind of hurdles they faced in terms of recovery? How the discriminatory patriarchal conditions pushed them to the verge of vulnerability? And what kind of social evils they witnessed? The interview was recorded, as women has a long list of complaints about their family, neighbourhood as well as the government. Mrs. Neelavathi Sekar from Rudharpuram shared her experience of surviving the unexpected situation. Their place was flooded twice, and the second time pushed them to relief camps for more than two weeks. With her two children, one boy and one girl, Neelavathi stayed in the nearby government school. She said, 'Food and blankets were provided by the volunteers. But overcrowding of relief camps wasn't easy to handle. Very few toilets were available and they turned to an unusable condition. I faced difficulties in getting medicines and sanitary napkins. Her husband works as a contractual labour and he left them in school and went out seeking money and relief materials. The children's school bags, books, and clothes were washed away. All they could save is their ID cards, ration card, and the little money they owned. The government provided ₹ 5000 as relief fund which came in handy but my husband's share was dumped in the liquor shop', she ended this on a sarcastic note.

Mrs Sasikala Sasikumar aged 28 is a domestic labour. She works for five families and her work starts from morning 5 am and she returns home at 7 pm. She is staying in Meenambalpuram and has two boys. Her husband works as a cab driver.

> Our livelihood was shaken by this disaster. On 1st December morning, it was a heavy down pour and the water level kept rising and entered our house. My husband was out for work and with two children I couldn't make any immediate decision. Unlike Vardah cyclone alert, we didn't get any prior warning through television or radio for floods. In spite of heavy rain, we suffered from drinking water shortage within the year.

She also condemned the government for having excessive Tamil Nadu State Marketing Corporation (TASMAC) shops.

Mrs Gomathi Karthik is 30 years old. She lives in Shanmugam Street. She got married when she was 16 and has two boys and one girl. She works as a sweeper on contract basis at the nearby shopping mall. On a light note, she shared her experience about meeting film celebrities who came for relief work. She said that most of the organizations and volunteers didn't reveal their identity or seek publicity. 'There were certain nearby places to pick up the relief materials. It was crowded by people and the materials were picked by the same persons again and again', she noticed.

Selatterpuram resident Kumari Suresh, aged 27, has two children, and she moved to her sister's place at Ambattur. She shared her experience working with a group of women in relief work. She said,

> I left my kids at my sister's place. There were twelve women and we cooked food for around two hundred people. My sister and other live in three floored apartments for rent. We got permission from the owner and used the terrace for cooking. We took the food in the neighboring streets and distributed to the people who were stuck without electricity, utensils or vegetables to cook.

However, in her opinion ₹ 5000 was not enough for relief as they lost their fridge and television and their vehicle was wrecked. Her husband is an auto driver and he couldn't earn for the next one month and they were left at the mercy of charity organizations and the government.

Kuruvamma Paneer, aged 58, live with her 70-year-old husband. Their two boys and two girls are married and are residing separately. In the times of floods, their age factor added to their constraints. 'Due to our old age, we were unable to respond immediately to the emergency. My children are living separately and due to the breakdown of communicating networks, we couldn't convey our situation to our children or relatives. After 12 hours of misery, we were rescued by the volunteers and left for the nearby safe place'. They live in the housing board at Egambaram Street.

Bommarani Rajini is 27-year old and lives in Arunthathiyar colony. The word 'Arunthathiyar' refers to a caste in which people are mostly scavengers. Most of the people are migrants from Andhra Pradesh. Her husband was working with the missionaries. They said that plenty of food was available but they were in short of volunteers to reach the food to needy. She said, 'I have a 10-month baby and I was really worried about the health of our baby girl. As the price of the milk hiked to ₹100 to 150 it was hard for us to afford'.

Liquor and Women's Anxiety

During the interview schedule and interaction with the respondents, it is widely noted that majority of male members used their relief fund to buy liquor. Many women complain that they couldn't tolerate the money being spent on liquor, while their family were suffering to afford for their livelihood. It also leads to health issues and they are subjected to domestic violence. Availability of liquor shops created some serious trouble in the scenario. While the city was in the midst of devastation, the NDRF and volunteers were struggling to reach certain areas, price of basic amenities such as milk, water, blankets, torch and batteries were hiked and hard to reach, the TASMAC has been functioning from 11 am to 11 pm. This issue was seriously expressed in social media. Consequently, public interest litigation was filed by Advocate Kanimozhi Madhi. According to the petition, 'Chennai has been facing rain and flood since November and during rescue and relief work. Closing all TASMAC shops in all the flood affected districts for 60 days will significantly bring down the menace of inebriation, lumpenism, and anti-social behaviour'. Hence TASMAC outlets were closed for 60 days in Chennai, Kancheepuram, Thiruvallur and Cuddalore district citing the experiences of counsellors, the petitioners said that during the times of serious loss, stress and trauma, a person might tend to seek escapist habits such as consuming alcohol. So, until life returns to the semblance of normalcy, it is advisable to make alcohol unavailable in the market as an effective remedy to prevent the despondent citizens from resorting to drinks. The PIL was taken to hearing by Chief Justice Sanjay Kishan Kaul and Justice Pushpa Sathyanarayana. The court declined the petition saying that this issue lies in the policy domain of the government (*The Times of India* 2015).

Consuming alcohol has an enormous impact on the occurrence and vigorousness of domestic violence. Poor economic status, social pressure, media and psychological stress lead to the consumption which has the high risk of conflict. World Health Organization (WHO) has been constantly stressing on this issue as it is strongly connected with the poor health of women. Studies were conducted which states that liquor directly affects the physical as well as cognitive function. It reduces self-control and pushes individuals to violence. Second, it creates financial difficulties and affects childcare. There are high risks of infertility. Third, marital relationships are jeopardized and lead to separation. Fourth, alcohol is used as an excuse for violent behaviour and in some cases, marital problems are induced to consume alcohol. Finally, the children who witness the domestic violence are likely to develop harmful drinking problems in their later stages (WHO 2006). Very few researches are pursued in connection with natural disaster, excessive alcohol consumption and domestic violence on women. Studies say the frequency of domestic violence against women increases during the disaster when the person starts to lose his control over physical health or wage-earning power. Such socio-economic disruptions collapse the social structure which regulates traditional behavioural norms. The non-functioning of formal as well as informal social control fails to resolve the issue (Minakshi 2013). Hence, the non-profit organizations and women's movements should see that they interact with the women and address the issues faced by them in the households. It should be discussed in the public forums and awareness should be created among women on responding to the crimes faced in the private sphere. Orientation programmes on laws related to women's rights should be conducted and the people should be trained to handle the legal proceedings. Second, the government should create employment opportunities for the economically weak section. It can also insist private sector to do so. By keeping them occupied with financial support, the crime rate can be reduced and the speed of recovery from the disaster may be quick. Finally, instead of targeting high income, the government should give priority to people's welfare and keep the liquor shops closed for a certain period during the disturbed situations.

Women's Representation and Participation in Disaster Management

Patriarchal dominance had been implanted in an institutionalized structure which forms a basic governing norm for our lives and we adhere to it. The state can be masculinist without intentionally or overtly pursuing the 'interests' of men precisely because the multiple dimensions of socially constructed masculinity have historically shaped the multiple modes of power circulating through the domain called the state. Hence, there is no intentional patriarchal imposition but the system gets sustained by conventional notions. For instance, instead of having 'active political subjects', the majority of women were treated as regulated, subordinated and disciplined state subjects. The same condition gets reflected in the process of disaster management. Women are constructed as citizens who seek protection. This notion increases the risks during past disasters. For instance, on 26 December 2004 Tsunami swallowed almost 12,000 lives. In the affected areas such as Nagapattinam in Tamil Nadu, about 2,406 women died compare to 1,883 men. Many women sacrificed their lives trying to protect their children. Significantly many women (most from fisherman community) died just because they don't know how to swim. Based on the conventional norms, women from fisherman community are prevented from entering the sea. Thus, tsunami revealed the women's poor potential and dependency on men. Most of the Indian women irrespective of rural or urban areas don't know how to swim. Because of cultural clothing restrictions, women don't learn swimming. It created a massive impact during flood or tsunami and left them at the mercy of men. In the case of Chennai floods, the role of women in rescuing process is very low. On the other hand, women who stepped out to help faced different sets of problems. Many women volunteers were subjected to harassment during their relief work. Most of the credits go to the government who kept the TASMAC open for the survivors to relax and indulge in rogue activities. On the other hand, women volunteers shared their contact details through social media in order to reach out and help in the emergency. But, many of them ended up receiving inappropriate calls and texts from unknown men. They shared their experience in social media criticizing these men as 'perverts' misuse the situation to flirt with women (Bobins 2015). The state should take it into account to have strict law and order and punish the womanizers and create a conducive environment for women to involve in these activities. It is the responsibility of both state and the citizens to overcome the misogynistic attitude that undermines women's potential. Women, instead of remaining as a passive recipient, should overcome these barriers and actively participate in the rebuilding process. On the whole, they should be equipped with the skills to tackle unexpected hazards.

Preparedness and Effective Warning System

Sources say that the meteorological department issued the forecast outlook for 2015 Northeast Monsoon season and predicted the record breaking rainfall. But the authorities failed to react and reach the message to public (*The Hindu* 2016). When the authorities decided to let water from Chembarambakkam Lake to Adyar Lake on 1 December, there was a breakdown in the communication system. A respondent from Ambedkar Bridge residents also complained that they did not get any intimation regarding flood alert. Though they had access to media, they didn't get any news as such. As most of the male members left the home for work, it was difficult for women to take decisions. The UNDP addresses these issues and suggests the state to have an effective communication system. International conference was conducted on Disaster Risk Management in 2005, 168 countries endorsed the Hyogo Framework for Action 2005–2015. The research says that the early warning system may fail to reach women particularly who belong to the lower income group. Hence, it published a guideline for developing gender responsive early warning systems. In 2008, India and the UNDP coproduced the compilation of best practices titled 'Women as Equal Partners: Gender Dimensions of Disaster Management Programme'. The prior objective of this project is to support preparation of participatory multi-hazard preparedness plan through the preparation of response and mitigation plans for disaster risk management at state, district, block and village and ward level in 176 most multi-hazard prone districts of 17 selected states roughly covering 30 per cent of the population. Accordingly, the first training session was held in Nagapattinam district of Tamil Nadu. Women from SHGs were involved in the risk management. After the training programme, women gave a positive feedback. Gaining more information and skill, they feel more equipped. By playing a crucial part of the process, they acquire the sense of adequacy and contentment. The UNDP considers self-help groups as the best local institution to introduce new initiatives and facilitates the process of creating an alternative structure as well as gaining required acceptability by the community. Moreover, the involvement of women in formulating programmes will create a holistic approach to access the community needs. The second session was held in Thiruvallur district of Tamil Nadu. A puppet show on disaster risk management was conducted in a girls high school. The puppet play was about a woman who lives at the river bank, taking preventive measure to safeguard her house from any hydro-meteorological disasters. The show successfully reached the audience and imparted them the importance of gender mainstreaming in disaster management. The puppet show was conducted in 41 schools covering about 75,000 audiences. The UNDP suggested schools to have a committee and include girl students as a part of it to ensure equal participation in the activities (UNDP DRM 2008). It is evident that international organisation has set up the blueprint of information and guidelines for preparing women to face and recover from hazards. But it is the responsibility of the government to implement the policies effectively and prepare the people to recover from the state of vulnerability.

Discussion

As Chennai faces back to back disasters in December 2015 and December 2016, it would be unfair to conclude the chapter without discussing the 'cyclone *Vardah*' of 2016. Unlike the Asian tsunami in 2004 and the floods in 2015, Chennai was well prepared and faced 'cyclone *Vardah*' on 12 December 2016. The government issued an advisory for the people on 11 December 2016. The people were asked to stay indoors and not to let their children out. Holiday was announced for schools and colleges, and government issued a notification to private companies to grant leave or work at home for the employees. The public was advised to stock food, drinking water, medicine and keep their mobiles and torches charged. The army, navy and coast guards were instructed to be prepared for the rescue operation. The NDRF teams and fire rescue services were positioned in low-lying areas. The people were moved to the safe places. The government's public relation officer (PRO) was updating the relief and rescue work done by sending pictures. It was evident that the government has improved in terms of efficiency in managing natural disasters.

Nevertheless, the state should take off from this positive note and run a gender sensitive management. Women should be a part of the decision-making process. It is important to note the marginal participation of women in disaster management in the state as well as the non-state arena. The feminist critique explains that the patriarchy is in-built within the state's administration. Nancy Fraser's (1987) critical discourse oriented approach guides feminist to raise question on welfare policies where social inequalities are defined and addressed. She offers two major arguments to intervene and challenge the policy framework. First is the structural, which means the welfare measures for women are constructed to sustain the existing patriarchal structure. Though it benefits women, it reinforces poverty. The second issue is ideological, which means the welfare system curtails the space to debate and redefine what various groups of women need and whose need should be given priority. In framing policies, the needs of women are exempted from debates and taken for granted (Fraser 1987: 149). Brown (1992) adopted the notion of reform politics, which seek state's intervention in familial and sexual issues. The patriarchy has been sustained with the politics of protection and the politics of state-centred social policies. The 'politics of protection' forces women to be legitimately excluded from certain spheres in the name of protecting them from men as well as from the binary labelling such as dark and light; wives and prostitutes, bad girl or good ones (Brown, 1992: 14). The stigmatizing stereotyping women using cultural norms form the base for their vulnerable position. For example, women were enquired about their swimming skills during the interview schedule. Hardly few women know how to swim. Most of the women expressed that they did not have access to learning swimming as well as defence skills. The major reason behind this scenario is the conventional patriarchal attitude and 'politics of protection', which restricted women from certain spheres. These aspects leave women in the receiving ends and in the hazard situations, their lives and livelihoods are at high risk. Taking it as a cue, organizations should maintain data on women's swimming and self-defence skills. Training sessions should be conducted regularly in order to make women more empowered.

The second important aspect which was observed during the interview schedule is the lack of legal awareness among the women. They do not know how to legally challenge the violence they face at home as well as at the workplace. This is because the spheres of family, civil society and state are isolated from each other. The family is perceived as naturally constructed and the role of women as household labour or as a citizen is not acknowledged, and thus it depoliticizes and marginalizes their participation in civil society and state. On the other hand, the private sphere which is meant for women does not give them the voice against violence such as marital rape, wife battering and child abuse. These are not considered violence and fail to recognize the personhood inside the household. Hence, it is important for the upcoming feminist jurisprudence to acknowledge such issues. Programmes on gender sensitization and legal awareness should aim to socialize women as 'right bearers'. They should be equipped to avail their rights and the opportunities provided by the government. It is the responsibility of the government to set a platform for women and ensure a substantial representation and participation in the administration.

Third is regarding the availability of excessive liquor even during the hazardous situation. The case studies exposed the facts that the alcohol consumption plays a part in inducing violence against women. However, anti-liquor campaign across the state convinced the government to close about 1,000 liquor shops. Following that, the order issued by Supreme Court led to the shutdown of more than 2,000 liquor shops situated at the national as well as state highways. The order may be a relief for a while, but still the debates are going on for relocating those shops. Women's rigorous agitation against liquor should be taken into consideration. It is high time for the state to prefer welfare of the citizens over the high revenue.

The gender mainstreaming in disaster management should be seen through the lens of human rights framework. Any nation which claims itself as welfare state should ensure the citizens the right to life and personal integrity; the right not to be subject to ill-treatment; equal protection in terms of nation's external or internal conflict and hazards; and safety and security and individual liberty. Apart from pressures given by international conventions, as a democratic state, India should ensure gender sensitivity in all its agencies such as legislature, executive and judiciary and aim to achieve the status of what we call it as welfare state.

Note

1. http://moud.gov.in/upload/uploadfiles/files/RAYGuidelines.pdf

References

Bobins, Abraham. (2015). "Shocking! Women Volunteers Involved in Chennai Flood Relief Face Harassment", *News*. Available at http://www.indiatimes.com/news/india/shocking-women-volunteers-involved-in-chennai-flood-relief-face-harassment-248312.html

Brown, W. (1992). "Finding the Man in the State", *Feminist Studies,* Vol. 18, No. 1, pp. 7–34.

CARE. (2006)."Poor Women in Urban India: Issues and Strategies" (2006). CARE. https://www.careindia.org/wp-content/uploads/2017/05/Poor-Women-in-Urban-in-India.pdf

Coelho, K. (2016). "Placing the Poor in the Flood Path: Post-Disaster Slum, Resettlement in Chennai" *The Caravan*. Available at http://www.caravanmagazine.in/vantage/placing-the-poor-in-the-flood-pathpost-disaster-slum-resettlement-in-chennai

Draft Report (2014). "Slum Free City Plan of Action-Chennai City Corporation", Rajiv Awas Yojna, Tamil Nadu Slum Clearance Board, Chennai.

Fraser, N. (1987). "Women, Welfare and Politics of Need Interpretation", in *Unruly Practises: Power Discourse, and Gender in Contemporary Social Theory*. Minneapolis: University of Minnesota Press. pp. 144–160.

Government of India (2005). "National Policy on Disaster Management, 2005". New Delhi: Government of India.

Government of Tamil Nadu (2015). "Flood Relief". Available at http://www.floodrelief.tn.gov.in.

Kolappan, B. (2015). "18 Patients Die in Chennai Hospital". *The Hindu*. Available at http://www.thehindu.com/news/cities/chennai/14-patients-die-in-private-hospital-in-chennai/article7948721.ece

Minakshi, Joshi. (2013). "Disaster and Domestic Violence: The Alcohol Connection", *Man in India*, pp. 551–559.

Ramya, Kannan. (2015). "Displaced Families Count Their Losses". *The Hindu*. Available at http://www.thehindu.com/news/cities/chennai/chennai-floods-aftermath-displaced-families-count-their-losses/article7958695.ece

Relief Web (2015). "Preliminary Appeal: Flooding in Tamil Nadu, South India—IND152 Appeal Target: US$ 657,278" (2015). Relief Web. Available at http://reliefweb.int/report/india/act-preliminary-appeal-flooding-tamil-nadu-south-india-ind152-appeal-target-us-657278

Ross, Tom and Lott, Neal (2003). *Billion Dollar Climate and Weather Disasters 1980-2003*. National Climate Data Centre, US Department of Commerce, Asheville.

The Hindu (2016). "Cyclone Vardah Brings Chennai to a Standstill". Available at http://www.thehindu.com/news/cities/chennai/Cyclone-Vardah-brings-Chennai-to-a-standstill/article16798323.ece

The Times of India (2013). "Natural Calamity Prone States Have No Disaster Management Mechanism: PIL". Available at http://timesofindia.indiatimes.com/india/Natural-calamity-prone-states-have-no-disaster-management-mechanism-PIL/articleshow/21177283.cms

The Times of India (2015). "Chennai Floods: Madras HC Refuses to Order Closure of Liquor Shops". Available at http://timesofindia.indiatimes.com/city/chennai/Chennai-floods-Madras-HC-refuses-to-order-closure-of-liquor-shops/articleshow/50171640.cms

UN/ISDR (2001). (United Nations International Strategy for Disaster Reduction) "Women's Human Rights in Disaster Context: How Can CEDAW help?" Division for the Advancement for Women (DAW), Turkey: United Nations.

UNDP – DRM (United Nations Development Programme – Disaster Risk Management) (2008). "Women as Equal Partners: Gender Dimension of Disaster Risk Management Programme". UNDP – DRM: Government of India.

WHO (2006) "WHO Facts on Intimate Partner Violence and Alcohol" (2006). John Moores University, Centre for Public Health, Switzerland: WHO.

11

Indigenous Knowledge System and Disaster Management among the Indigenous People of North-eastern Himalayas, India

Ajeet Jaiswal

Introduction

Global concern for finding appropriate indicators sterns partly from our inability to interpret early warning signals of major ecological changes, given the complexity of ecological system. But this concern also stems from a faith that a large number of communities around the world have survived precisely through interpretation of such signals for various short-term and long-term adaptation (Gupta and Suther 1996).

In India, local communities had well-developed traditional indigenous knowledge systems for environmental management and coping strategies, making them more resilient to environmental change. This knowledge had, and still has, a high degree of acceptability amongst the majority of populations in which it has been preserved. These communities can easily identify with this knowledge and it facilitates their understanding of certain modern scientific concepts for environmental management including disaster prevention, preparedness, response and mitigation. Indigenous knowledge is a precious national resource that can facilitate the process of disaster prevention, preparedness and response in cost-effective, participatory and sustainable ways. Hence, a blend of approaches and methods from science and technology and from traditional knowledge opens avenues towards better disaster prevention, preparedness, response and mitigation. Globally, there is increasing acknowledgement of the relevance of indigenous knowledge as an invaluable and underused knowledge reservoir, which presents developing countries, particularly India, with a powerful asset in environmental conservation and natural disaster management. Specifically, from time immemorial, natural disaster management in Africa has been deeply rooted in local communities which apply and use indigenous knowledge to master and monitor climate and other natural systems and establish early warning indicators for their own benefit and future generations.

Comparison between Western knowledge systems and indigenous knowledge systems has not explored the epistemological foundations of the latter (Shankar 2007). As a result, there has been a widespread monolithic approach to knowledge and learning. Thankfully, in India and elsewhere, traditional knowledge systems of local and indigenous communities has increasingly renewed a respect. Indigenous knowledge is being recognized as at par with and complementary to scientific knowledge. More and more researchers are turning towards traditional knowledge systems, not only to fill gaps in the existing knowledge but also to 'discover' novel knowledge concepts and viewpoints. The value of diversity of knowledge systems, therefore, has redefined the human quest for truth. Today, the orally transmitted and experientially based knowledge of local and indigenous people compares well against the experimental science of the technological era. Since tribal groups have lived within their local environments since time immemorial, they possess a rich knowledge of nature and natural phenomena. Hence, current environmental research can be enriched through exploring the traditional knowledge systems of such groups of people.

The term 'disaster' is difficult to define, but despite the 'blurred edges' of the definition, it is still a useful concept (Oliver and Hoffman 1999). A disaster can be regarded as a severe, relatively sudden and unexpected disruption of normal structural arrangements within a social system over which the system has no firm control (Nashreen 2004). It is also defined as a crisis situation that outstrips the capacity of a society to cope with (Anderson and Woodrow 1989). A disaster occurs when a significant number of vulnerable people experience a hazard and suffer severe damage and/or disruption of their livelihood systems in such a way that recovery is unlikely without external aid (Blaikie et al. 1994). It is argued that while hazards are natural, disasters are not. Social processes generate unequal exposure to risk by making some people more prone to disaster than others, and these inequalities are largely a function of the power relations operative in a society.

Critical to discerning the nature of disasters, then, is an appreciation of the ways in which human systems place people at risk in relation to each other and to their environment—a relationship that can best be understood in terms of an individual's, a household's, a community's or a society's vulnerability (Hilhorst and Bankhoff 2004). Disasters could be natural or human caused. In the former category are geological hazards (earthquakes, tsunamis/seismic sea waves, volcanic eruptions, mud flows), atmospheric hazards (hurricanes, tropical cyclones or typhoons, tornadoes, lightning-caused fires) and hydrological hazards (e.g., river or coastal floods). Human-caused disasters include heavy environmental pollution, industrial and technological disasters, major traffic accidents, epidemics and many fires. Sometimes disasters can be triggered by combinations of human activity and major natural disturbance. Disasters can also be acute (e.g., volcanoes eruptions, earthquakes, tsunamis, floods, hurricanes) or chronic (e.g., drought, famines, environmental breakdowns, pollution, toxic exposures) and can be focused in geographic scale or widespread. Whatever the cause or extent, coping with disasters is an ongoing human endeavour, obviously related to our collective survival.

Environmental conservation and natural disaster management are important in the livelihoods of indigenous people who often live in hazard-prone areas and have built up, through thousands of years of experience and intimate contact with the environment, a vast body of knowledge on disastrous events. This knowledge is a precious resource that continues to contribute to environmental conservation and natural disaster management in these regions (UNEP 2006). In this study, we focus on documentation of the traditional knowledge of the Lepcha people of Sikkim vis-à-vis disaster management and mitigation.

Sikkim is the second smallest state of India, located in Himalayan range between 275 N to 289 N and 8759 E to 8856 E, with an area of 7,096 square kilometre. The Lepcha people of Sikkim are a cultured people, with their own particular knowledge system, worldview and oral history. Despite the underlying similarity of tribal ethos they share with the numerous tribes of Northeast India, their cultural and religious beliefs have a distinctive character. Notably, the oldest surviving names of various mountains, hills, gorges, rivers, lakes and caves in Sikkim are Lepcha in origin.

These people are mostly Buddhist (a religion embraced in fifteenth century) but many of them have now adopted Christianity (embraced in nineteenth to twentieth century). Lepcha folklore is rich with elaborate narratives, which are often a reflection of scientific reality. Their language and biological classification systems are also unique. Almost all of the Lepcha names of animals start with the letter 'Sa', and the names of different types of snakes and various bamboo products start with the letters 'Ta' and 'Pa', respectively. Hence, surya, suthong, su-chyak, suna, suko, sangi, sattim and sahu are the names of tiger, leopard, bear, dear, lion, porcupine and monkey, respectively. Panu-bu, pamol-bu, panul-bu and palong-bu are some of the names for various kinds of snakes. Similarly, ta-zyang, tangar, tangjung, tafu, takchim and talyung are the names of different bamboo products. The Lepcha also have an extraordinary understanding of birds and bird behaviour (Jha 1995).

The rationale of the study is that it is important to document this knowledge as it highlights the complex traditional knowledge of an important group of people whose contribution to the general knowledge base has not been well recognized or appreciated. Present chapter has high significance because such traditional knowledge is being lost at a rapid pace. Traditions such as storytelling are disappearing altogether, making their documentation particularly crucial. Traditional knowledge can be a useful input in our endeavour to face the challenges posed by disasters. It is also expected to be more effective as it has been consolidated through a process of sociocultural evolution.

Methodology

The study is based on general observations and on general secondary information collected from different manuscripts and sources. The study area pertains to Sikkim in general. The study is largely exploratory in design and is based on literature reviews. Secondary data was collected about the indigenous Lepcha people and from Indigenous Lepcha Tribal Association report.

As no apparent formal or structured traditional knowledge exists on disaster management amongst the Lepcha people, the researchers attempted to put the information collected into right context. Rather than being quantified, the data and information are presented in text form for easier understanding.

Data of the present chapter discusses about anthropological and sociophilosophical interpretation of disasters, preparedness, effect of disasters and responses to them and mitigation.

Results and Discussion

Anthropological and Sociophilosophical Interpretation of Disasters

Anthropological and philosophical aspects of a natural phenomenon assume importance in the context of disaster management (Bankhoff 2003; Sen 1981). The perceived cause of a disaster is an important anthropological and philosophical inquiry if an effective management plan is to be envisaged. The Lepcha people have no real concept of sin. When an individual does something terrible, such as indulging in incest, fraud, murder, violating accepted norms of hunting/food-gathering or showing disrespect, he/she invites no individual divine punishment. Such acts by individuals produce horror rather than revulsion from others because each one could result in a whole year of disaster and collective suffering by the entire village via a series of divine acts reflected as a natural calamity. This anthropological and sociophilosophical construct is known as *nam-toak*.

Disasters in Lepcha culture, therefore, have been traced back to evil intents of human beings. Such an interpretation demands that individuals always behave appropriately and responsibly to avoid causing disasters. Leaving aside the scientific veracity of such a belief, it can still be argued that this approach helps to promote a strong social fabric and effectively neutralizes the selfish, malicious and wicked impulses of individuals. It fosters a spirit of camaraderie, unity and harmony that is so essential for disaster management. It is perhaps not an accident that there is no historical instance of Lepchas invading other lands. Even when others began to settle down in their territory through invasion or migration, there have been hardly any instances depicting Lepcha aggression, although bitterness at losing precious resources to others is visible.

It is also important to determine whether such a philosophy can be validated through any underlying principle. Do our greed, jealousy, maliciousness, arrogance, fear, hatred and other evil intents cause disasters? They do. Mahatma Gandhi once observed, in the context of European colonization, that the earth has enough for our need but not for our greed. Thus, greed can lead to disaster—especially human-caused environmental disasters. Terrible wars have been fought because of fear, hatred, greed and arrogance. Noted economist and Nobel Laureate Amartya Sen holds the view that there was actually no overall shortage of rice in Bengal in 1943 when an estimated 3 million people died of famine. He showed empirically that availability of rice was actually slightly higher than in 1941, when there was no famine (Sen 1981). Greed, corruption and exploitation not only result in disasters but also seriously compromise our ability to manage disasters. Most disaster management efforts remain just pretence because of inherent human weaknesses. Hence, their relationship with disaster is not a figment of imagination but a sad reality.

There are several examples of how corruption can harm those affected by disaster, including aid workers exchanging food for sex in West Africa, survivors of the tsunami in Aceh losing their homes when contractors built them without foundations and village chiefs diverting food from the most vulnerable in India in 2001. In 2010, in Pakistan the greatest challenge to the flood relief that hit the country was posed by the prevailing corruption in that beleaguered country. Is there any Lepcha knowledge that supports such an interpretation? According to one of the traditional stories (Gorer 1996), once there lived seven brothers with seven wives, seven daughters, seven sons, seven daughters-in-law and seven sons-in-law. They had to work hard day in and day out. Yet their widowed sister, along with one daughter, lived in prosperity not far away. The brothers and their wives grew very jealous and decided to kill the sister and her daughter and take away their property. As per their plans one day the brothers and their wives attacked the helpless sister and cut her into pieces.

The sister, however, was a powerful *Mun* (priestess), and hence she was able to rearrange all her chopped-up parts into a coherent whole once again. She told her daughter everything and advised her to not to resist her maternal uncles when they came to take away their property. The uncles did just that, leaving only a hen and a calf for the daughter. The dead sister, meanwhile, went and entered a cliff and prayed to divine beings for revenge. Her prayers were answered. Seven days later a handsome man came and married her daughter, and they began to live happily. For the seven brothers and their families, however, they followed a year of disaster and all but two sisters-in-law died. Even these women suffered from disease and soon they too died. Since then, the Lepchas believe that one should not show off his/her or wealth as otherwise *Ginoo mung* (the ghosts of the seven brothers and their families), invoked by the envy/jealousy of others, would play havoc in their lives. *Ginoo mung* is an incarnation of other people's envious thoughts. Similarly, the jealous and malicious ones also go through a year of disaster as a divine punishment for their wicked thoughts.

According to one of the versions of this story encountered, once the invocation by the dead sister inside the cliff was over, an earthquake ensued, destroying the evil brothers and their village. Thus, earthquakes too are believed by the Lepcha people to be caused by evil acts of human beings.

In yet another story (Jha and Jha 2008), once the Lepchas living in Sikkim were attacked by invaders from Bhutan. They fought back bravely. The war lasted longer than 12 years and as a result agricultural chores were neglected. This resulted in a prolonged famine. There was no respite and people had to travel long distances to find meagre food. One day a boy and his grandfather were searching for food, but they found nothing. At last, the grandfather requested a bird to help them. The bird took them to a tree that has a large honeycomb. The boy was happy but the grandfather said that the honey on that hill was poisonous. Both were disappointed until a bear came and taught them how to neutralize the poison and make honey edible. This story establishes that famine occurred primarily because of wars and not as a natural calamity. This folklore also teaches us the principles of resilience, diversification of our resources and behavioural modification to seek help in the face of disaster.

Preparedness

It is expected that traditional knowledge systems of indigenous people of any region will have some distinct elements relating to pre-disaster preparedness. Awareness of disasters is the first step in preparedness. Natural disasters are often region specific. A particular region generally has a definite set of natural disasters occurring within the area. Which natural disasters generally occur within the geographical boundaries of Sikkim Himalayas? Lepcha's traditional memories of disasters (as recounted in their folklore) are of flash floods, earthquakes, forest fires, landslides, draught, famine, high-velocity winds, hailstorms and avalanches.

Researcher has already narrated the story depicting famine as one of the disasters Lepchas have faced. They also have a very strong memory of flash floods as is told in the story, *The Great Flood*. According to this story, the two major rivers of Sikkim—Rungeet and Teesta—happened to be husband and wife who in the ancient times had eloped to get married. Once it so happened that because of some silly misunderstanding the husband and the wife had a big quarrel. To outdo each other in fury they began to swell beyond themselves. The water in the rivers began to rise higher and higher. Soon the entire Nye-ma-yel (Sikkim) was flooded. All the animals and human beings cried for help, requesting the husband and the wife to have pity on them and to stop quarrelling. But they paid no attention to the distress they were causing to others and kept on pouring their wrath on each other. Soon the situation became impossible. Wherever, one looked, only water could be seen. Not only the villages situated in the lush green valleys, but those located over mountain tops too were now under water. The animals and the common folk panicked and ran towards the crest of the tall mountains. The rivers kept on swelling, and the water continued to rise and at last only mounts Tendong and Montom stood above water. These two peaks were brother and sister. Both were unhappy about the plight of innocent lives and took it upon themselves to protect the poor animals and the human kind from the fury of the quarrelling couple.

But the water was still rising. To keep them safe, the brother and sister rose higher and higher above the water. It was a difficult feat; both were terribly exhausted and grew weaker with every passing moment. Only the welfare of the animals and people kept them going. But how long could this continue? The sister, unable to endure any more, sank under the water even as she prayed, frantically calling out to the gods to protect her dear brother. All the people and animals who had taken shelter with her thus went under water and died. The death of his sister came as a great shock to Mount Tendong.

Yet, he could not afford to shed tears as that would have increased the flood. With great effort he held back his tears and kept rising higher and higher. Those who had sought shelter on him were sure that they were doomed. The fate of those who died with Montom made them more frightened than ever. They prayed with all the sincerity, invoking the Goddess Na-zong-nyo to come and protect them. Na-zong-nyo too was distressed by what she saw. She knew there was no point in trying to restrain the fighting couple who obviously were beyond any reason now.

She called the blood pheasant and asked him to go and protect them. The blood pheasant flew down to the Mount Tendong and perched on its peak. All eyes looked towards him with hope. He thought for a while. Then, he accepted *chi* (a local drink) from those trapped over the mountain and drank it. The *chi* made him thirsty as never before. He dipped its beak in the floodwater and began to drink it up in great gulps. Soon the water began to recede. Everyone heaved a sigh of relief. The water went down further and not much later other mountains, lush green valleys and the Lepcha villages and homes that were submerged emerged. It was only after seven days that the blood pheasant stopped drinking. By now it was normal everywhere. Teesta and Rungeet had stopped quarrelling and were ashamed of their foolish behaviour. The story underlines another vital disaster management tenet—It is up to the individual to know what to do in an emergency.

Story of the Utis Tree (Himalayan alder) that recounts later depicts that Lepchas have been familiar with landslides since time immemorial. Forest fires have also been mentioned by Lepcha people again and again in discussion with them. Some of the fires, they say, are natural, whereas others are caused by people. The Lepchas have been practicing shifting cultivation for centuries, for which they clear new land by burning over a patch of forest. Sometimes these fires get out of control and this can lead to disaster. The first instance of forest fire appears in the Lepcha narrative about how fire was brought to the earth. According to this story (Doma 2010), Narip Nom asked Tarbong to bring gifts for her before she consented to marry him. This task was difficult as it required fire to be stolen from a *mung* (an evil supernatural being) called Deut mung who had hid the fire in a place called Mashyok Matel. Ka-krhyak fo (a black-backed Kaleej Pheasant) agreed to steal fire so that the marriage could be solemnized. He succeeded in stealing the fire but while he was returning he put the fire on a chestnut tree and went away to find food for himself. A strong wind blew and the fire spread in the forest. Ka-krhyak fo, unfortunately, was caught in it, and as a result he lost his wings and skin around his eyes had peeled off patches and were reddish.

The Lepcha also tell stories of earthquakes that rocked their land, which resulted in the destruction of property and lives. They also share their experiences and oral histories on avalanches and hailstorms. But we do not come across any Lepcha story or experience of tsunamis, volcanoes or hurricanes, as such natural disasters do not occur in Sikkim.

The Lepcha also possess knowledge of the annual timing around which such natural disasters are expected to occur. Thus, landslides and flash floods occur during the monsoon season (June to September), and forest fires in summer (April to May). A drought is related to rain-less winter and hailstorms happen in March. High velocity winds occur in April to May. One of the Lepcha songs (Jha and Jha 2008; Jha and Jha 2011) is as follows:

> The hills are silent,
> The valleys are hushed,
> The cuckoos are back to Mayel Place*
> It is the month of Num-kum**
> Rains are violent and incessant Time to harvest wet paddy.
> *Mayel Place, Paradise, **Num-kum (July-August)

Lepcha's traditional knowledge of disaster indicators is also substantial. Researcher has already mentioned that a rainless winter indicates local drought. Famine is indicated when bears (Ursus thibetanus) begin to raid villages, as well as when langurs (Presbytis entellus) move around in smaller groups. Both suggestions are logical because when there is not enough food in the forest, the bears will invade villages for food, and langurs benefit by moving around in smaller groups as this way they can cover far bigger areas and find sufficient food during scarcity. Lepchas living in the higher reaches believe that harsh weather and consequent famine are indicated well in advance by wolves (Canis lupus chanco). Wolves, according to them, do not breed at these times, in anticipation of such conditions. Such views are also reflected in Indigenous Wisdom of Siberia and Canada (Mowat 2010). Wolves, too, are extremely sensitive to the weather and environmental conditions and hence it is quite possible that they sense an impending shortage of food and accordingly, hormones controlling mating behaviour are not released into their system. This hypothesis, of course, needs to be corroborated.

The Lepcha believe that if the birds are silent, rain and storms are due. Alternately, if the birds are singing loudly and flocking together happily on the ground, good weather will continue. One of the common beliefs prevailing in Northeast India (The Telegraph India 2008), East Asia (Khonumthung News 2007), Bangladesh (AFP Bangladesh 2008), China (Mizzima News 2008) and also shared by the Lepcha (Tamsang 1998) is that in the year when particular bamboo species bloom en masse, a famine will occur. This is believed to happen every 48 years. The bamboo flowering cycle is called 'Mautam' by the Mizo people of Mizoram and the related Chin people of Burma. It is called 'Yu Li Hku' (rat famine) by the people of Kachin State in northern Burma. Such beliefs do have a scientific basis (rats and kyats) since, after blossoming, the bamboo produces fruit, then dies off. The fruit is a large seed, resembling an avocado, and is packed with protein and other nutrients. During the fruiting stage, local species of forest rats feed on the bamboo fruits. With this nutrient boost, the rats cease cannibalizing their young and begin to reproduce in an accelerated birth surge, producing a new rat generation as often as every three months. Once the burgeoning population of rats has stripped the forest of bamboo fruit/seeds, however, nocturnal rat swarms quietly invade farms and villages to devour crops and stored rice, maize and other grains, potatoes and other vegetables, chili and sesame. The rodents often grow to exceptionally large size and can grow through bamboo and wood floors, walls, storage containers and granaries.

It is not surprising that wild animals have evolved to respond to their natural environment, coping with seasonal flooding, wild fires, landslides and earthquakes. In Sri Lanka, Asian elephants carrying tourists on the beach ran away minutes before the tsunami struck. Many locals spoke of being saved as they chased their dogs, which were also fleeing from the wall of water. This response reduced losses (Dawn 2010). There are other such observations too (Kirschvink 2000; Quammen 1985; Schaal 1988).

The Lepcha people believe that the behaviour of various pheasants (cries and nervousness) as displayed right before a major earthquake is an important indicator. Such a belief prevails in China also. The most important earthquake bioindicators, according to the

Lepcha, are weasels, martens and shrews—animals with high metabolism and rapid heartbeat. Their unusual behaviour (panic and confused movements), in response to earthquakes, may be triggered by ground tilting, groundwater changes, electrical or magnetic field variations (Kirschvink 2000).

Forest fires are said to be indicated by sudden movements of birds. Such a belief is understandable as unlike earth-bound creatures it is the birds that will move out first. But as the actual time of forest fires is known to the Lepchas they keep vigil during this time, particularly at night when fires are easily visible from a great distance. Landslide-prone areas can be identified easily by a visual inspection, and Lepcha people know well to not to construct houses in landslide zones.

Lepchas hold an interesting view on the migration of birds. They believe (Jha 1999) that Mayel fo (migratory birds) bring messages sent by Mayel Gods to inform people of the right time to sow, weed and harvest crops. An approximate English translation of a Lepcha song (Siiger 1967) is as follows:

> Tsuk dun dun, Tsuk dun dun, Tsuk dun dun, Tuk bo jo, Tuk bo jo, Tuk bo jo Tuk fyel jo, Tuk fyel jo Tuk fyel jo, nam du tsat [The proper time to sow, Rice has arrived, It is Jing month (March-April) Let us sow, My family and friends, Let us join our hands for this task, If we do it will bring happiness for us].

Researcher come across names of three cuckoos in the song: Tsuk dun (Himalayan Cuckoo), Tuk bo (Indian Cuckoo) and Tuk fyel (Brain fever bird).

Similarly, the song for weeding is as follows (Jha and Jha 2011):

> Time is for new flowers to bloom, Time is for new leaves to sprout, Time is for shifting seasons now, for the rushing gales and rumbling thunder, Time is for weeding now, Cuckoo it is proper time to weed, The season has now come to us, Time to weed paddy Riao-Riao.

This song also mentions cuckoo, as well as an unknown bird called 'Riao-Riao'.

How is this tradition related to natural disasters? The Lepcha believe that any preponement or postponement in the arrival of Mayel fo (migrating birds) indicates the possibility of natural disasters such as hailstorms that can adversely affect agricultural activities.

Effect of disasters and responses to them: Disasters have seriously detrimental effects on living beings. They disrupt normal life, as depicted well in the story of the great flood recounted previously. Some groups of people, however, are more vulnerable than others. According to one Lepcha folktale (Jha and Jha 2011), once upon a time there were a brother and a sister. Unfortunately they were orphaned while they were still young. The sister, being the elder, went around in the wilderness gathering food for herself and her brother. Once it so happened that the sister found nothing to eat for many a days. She could see that her brother would soon die of starvation. Then, just to keep his hope alive, she pretended to have found some sweet potatoes and began to dig for them. It went on for a long while with the brother asking every now and then for kew (sweet potatoes) and sister replying kyon (Yes! I am giving). The sister kept on digging in despair but found nothing. The hole in the ground where she dug got bigger and bigger until its wall caved in and the unfortunate sister was buried alive. Meanwhile the brother, too, died. After their death, both were transformed into yongo-hu-pho (Sikkim Rusty-cheeked scimitar babbler). Even today, the Lepchas believe that the brother keeps on calling kew—asking for sweet potatoes from his sister and she keeps on replying kyon—'I am giving'.

Children, particularly the orphaned ones, are thus especially vulnerable to the painful effects of disasters. Even the story recounted previously of the Bhutanese invasion implies that the main sufferers were children and the aged. In yet another folktale (Jha and Jha 2011), once there lived a husband-and-wife pair of mati-tap-pho (white-capped redstart, Chaimarrornis leucocephalus). They were very fond of each other. One day the wife fell ill. The husband put her in bed and began completing the domestic chores. A little later he went down to the riverbank to wash clothes. While he was washing, there was a flash flood and he drowned in it.

Later the wife came out to look for her husband. She went near the river but by then the flash flood was over and everything appeared normal—there was no sign that her husband had been washed away. The wife began searching for her husband, calling out cee-bee, meaning where you are. She can still be seen flying alone near rivers and streams and crying out cee-bee! Lepchas call the bird

Cee-Bee-Pho alsom and insist that it is the only bird to move towards high altitudes even during the severest part of winter. This story indicates that those living or working in areas susceptible to disasters are more vulnerable than the others.

The story of great flood underlines yet another important issue. When a disaster strikes, common people are left to fend themselves with no hope for outside intervention. In this story, the Goddess intervened much later to save some of the Lepcha people. The story is repeated today as well. Governments and other agencies respond to disaster much later, even though most destruction and loss of lives takes place in the first few hours of a disaster. The common people, therefore, should be provided with knowledge, ability, skills and resources to take care of themselves in the face of disasters. The story also highlights that the Lepcha people knew that they had to climb high in the mountains to escape the great flood. Knowledge of safe areas during and just after disasters is crucial for survival.

Mitigation: Many strategies employed by the Lepcha to help them to overcome the after effects of a disaster can be identified. The people make their houses of timber (from a tree called sambrang kung) and bamboo. Such houses are known to be quite stable in the face of earthquakes. Even if such houses break, they do not usually result in the loss of lives. This could be one reason why earthquakes tend to cause greater damage in modern settlements with heavier house construction than in traditional Lepcha settlements.

Traditional Lepcha houses are built on rock and on the higher reaches, avoiding floods and flash floods. The floor of a Lepcha house is built up to a metre (2–3 feet) higher than the ground, with some space, called tanhanp, left below. This is done to protect the house from the turbulent rainwater that might flow beneath the dwelling. The houses, we have already indicated, are not built in the landslide-prone areas. Continual availability of food during and after a disaster remains an important aspect of mitigation. The Lepcha have evolved a sophisticated philosophy to engender conserving food. They show restraint in the killing of game or gathering of plant material protected by taboos, which are invariably a veneer over a practical measure. The Lepcha socio-ethical norm, called 'buk-rup', conveys a unique philosophy. It essentially involves avoidance of unnecessary wastage, destruction of nature and employment of simple techniques for its improvement. This entire belief is reflected in many Lepcha traditions.

Even the story where bear teaches people about detoxification of poisoned honey is not merely an idea. The Lepchas practiced it widely. The Lepcha elders embarked upon the process of neutralizing the toxic contents of otherwise edible items for their consumption as food. This required an investigative task of some magnitude. After many efforts, the Lepcha priests succeeded in converting many otherwise poisonous plants into harmless edible food sources. The philosophy and the technique of neutralizing the toxic content of edible plants for their consumption as food is called 'Sim-Anyu-Mun' (Jha 2001). This indicates that disasters often pose unique problems rarely faced in daily emergencies. Human response needs to be equally innovative.

Researcher come across another interesting folktale of Lepchas, in which a male Rhododendron tree once fell in love with a lady Utis tree (Himalayan alder) and approached to her father to communicate his intentions to marry her. Father Utis tree found his proposal ridiculous because Rhododendron tree was a dwarf and very shabby looking. He rebuked him and took a promise that he will never come up with such a proposal. A sad Rhododendron tree returned home. This happened in the autumn. A few months passed by and it was time for spring season. The father Utis tree recalled the ugly Rhododendron and laughingly told his daughter about the ridiculous offer made by the ugly and shabby Rhododendron. His daughter did not share his sentiments however and she went looking for the rejected suitor. There stood the Rhododendron tree in full bloom and looking heavenly. Lady Utis tree fell in love with him and begged him to marry her. The Rhododendron tree told her that as he has already made a promise to her father, now he cannot marry her. The lady Utis tree was crestfallen and in intense agony she jumped from the mountain and died. Since then, the Lepchas believe that the Utis tree grows in landslide-prone areas. Every time a Utis tree is swept down in a landslide, the Lepchas repeat this story.

This knowledge, reflected in the story, about which trees grow in landslide-prone areas, can be useful to control landslides by planting herbaceous plants (rather than hardwood trees) over fresh landslides. Once these plants take root and hold the topsoil, gradually hardwood plants can be planted to convert the landslide into a forested area.

The Lepcha also associate certain flora particularly with forest fires. This may have a scientific basis as certain plants capitalize lack of competition in forest areas that are burned over. The people also tell about plants that recover first after a fire. Ecologists call them 'resprouters'. Knowledge of such plants can consolidate efforts to mitigate the aftereffects of forest fires. Certain plants are known to be resilient to forest fires. These include mostly large trees whose inflammable crown are high up, far away from the effect of fires, and whose roots run deep inside the earth—again safe from fires. This type of knowledge of the Lepcha and other indigenous people needs to be documented in a structured manner and incorporated, with appropriate permission and benefits, into a region's general afforestation plans and policies.

Conclusion

Comparison between Western knowledge systems and indigenous knowledge systems has not explored the epistemological foundations of the latter. Traditional knowledge has only recently touched the scientific chord of Western academic knowledge. More and more researchers today are turning towards it not only to fill gaps in the existing societal knowledge but also to find novel knowledge concepts and viewpoints for application in a contemporary context. Since tribal groups have lived within their local environments since time immemorial it is obvious that they do possess a rich knowledge about nature. Hence, research on nature and environments is enriched by incorporating traditional knowledge of indigenous and tribal peoples. Such knowledge is a precious national resource that can facilitate the processes of disaster prevention, preparedness and response in cost-effective, participatory and sustainable ways. This chapter is a product of review of literature and it explores the traditional knowledge on disaster management of the tribal people of Sikkim. It establishes that in the guise of folklore, the tribal people of Sikkim have developed an elaborate understanding of the nature and causes of disasters and have identified accurate and precise indicators to assist in predicting disasters, as well as ways and means of mitigating their effects. Traditional knowledge of tribal people of Sikkim finds echoes in some of the current principles of disaster management. We conclude that traditional knowledge of tribal groups all over the world not only needs to be recognized, conserved and documented but also is to be incorporated into efforts to formulate effective disaster management strategies.

From the information presented here, it is apparent that the Lepcha people of Sikkim have experienced a range of disasters since the time immemorial. They have evolved elaborate philosophies to help explain the nature and causes of disasters. These beliefs, as reflected in their stories, appear folkloristic but they are based on sound principles that help to guide actions and behaviours that can help anticipate and limit the impacts of disasters. Also, these philosophies help people to develop close-knit social structures that make them more resilient in the event of a disaster. The importance of these social relationships and institutions can be expressed by the maxim that those who work together well on a daily basis tend to work together well at times of disaster.

Present study has high significance because such traditional knowledge is being lost at a rapid pace. Traditions such as storytelling are disappearing altogether, making their documentation particularly crucial. Traditional knowledge can be a useful input in our endeavour to face the challenges posed by disasters. It is also expected to be more effective as it has been consolidated through a process of sociocultural evolution. The Lepcha also possess detailed knowledge of their natural environments which makes them well aware of the potential disasters facing them as well as of the indicators that help provide sufficient warning. They have evolved various ways and means to avoid and escape unpleasant aspects of disasters. Even their seemingly compromising posture in relation to disasters could be rather interpreted as their silent courage to bear what nature offers them.

Their knowledge needs to be recognized and incorporated in our general efforts to formulate a comprehensive and holistic approach to disaster management in India and elsewhere in the world. It has been understood that disaster management should be based on what people are 'likely to do' rather than what they 'should do'. Traditional knowledge, therefore, will provide more effective measures in coping with, enduring and managing, disasters as they are 'known' and 'understood' by specific group of people. The knowledge gained from local community, as from Lepchas, could be great depositories to complement with the formal programmes of government. This kind of information could be of more use in making local plans for the region. The knowledge held by local community could either be blended or be grafted over scientific knowledge on climate change to mitigate the climate change/weather anomalies in order to mitigate the disasters.

Acknowledgement

Researcher would like to give special thanks to all the librarian and researcher of our country as much of the information provided in this chapter was collected from different library and website related to traditional knowledge. The researcher also gratefully acknowledges Vanya Jha and Ajeya Jha who have narrated the folktales, and apologies for not being able to acknowledge all of the people by their individual names. I am especially grateful for the assistance and encouragement of Prof. A. K. Kapoor, Prof. Satwanti Kapoor, Prof. T. S. Naidu and Prof. A. Chellaperumal.

References

AFP Bangladesh (2008). "Rat Invasion Fuels Famine Fears in Remote, February 9. http://afp.google.com/article/ALeqM5jkzeqnrBjs_BRfNVWbN1Lh2kld6g.

Bankhoff, Greg (2003). *Cultures of Disaster. Society and Natural Hazard in the Philippine*. London/New York: Routledge Curzon.

Blaikie, Piers, Cannon, Terry, Davis, Ian and Wisner, Ben (1994). *At Risk: Natural Hazards, People's Vulnerability, and Disasters*. London: Routledge.

Dawn Smith, M (2010). "How do Natural Disasters Affect Wildlife?" Bulletin of the Seismological Society of America, Vol. 98, pp. 430–443.

Doma Yishey (2010). *Legends of the Lepchas*, India: Tranquebar Press of Westland Ltd.

Gorer, G. (1996). *The Lepchas of Sikkim*. New Delhi: Gyan Publishing House.

Gupta, A. and Suther, J. (1996). "Indicators of Indigenous Ecological Knowledge: Lamppost, Cross Road and Turning Point". IIMA working paper. No. WP1996.

Hilhorst, Dorothea and Bankhoff, Greg (2004). "Introduction: Mapping Vulnerability", in *Mapping Vulnerability, Disasters, Development and People*, ed. Greg Bankhoff, Georg Frerks, and Dorothea Hilhorst. London: Earthscan.

Jha, A. (1995). "Ornithoscopic Norms of Lepchas of Sikkim", *Amrita Bazar Patrika, 11–19*.

Jha, A. (1999). "Nature Folktales from Sikkim and Arunachal Pradesh". World Wide Fund for Nature India. 23–33.

Jha, A. (2001). "Ecological Prudence of Lepchas of Sikkim", Tiger Paper 29, p. 1, FAO Bangkok.

Jha, V. and Jha, A. (2008). *Bird Behaviour as Depicted in Lepcha Folklores*. Paper presented at Pan-African Ornithological Congress in South Africa, September.

Jha, V. and Jha, A. (2011). "Traditional Knowledge on Disaster Management: A Preliminary Study of the Lepcha Community of Sikkim, India", *Indian Journal of Traditional Knowledge*, Vol. 10, No. 1, January, pp. 173–182.

Khonumthung News (2007). "Gigantic Rodents Attack Paddy Farm and Stocks". October 27. http://83.143.83.35:36080/VirtualHostBase/http/www.khonumthung.com:80/kng/VirtualHostRoot/kng-news/07-news-archive/oct-07 -news/ gigantic-rodents-attack-paddy-farm-and-stocks/

Kirschvink Joseph, L. (2000). "Earthquake Prediction by Animals: Evolution and Sensory Perception", *Bulletin of Seismic Society of America*, Vol. 90, pp. 312–323.

Mizzima News (2008). "Swarms of rats destroy crops in townships in Chin State" July 28, Zalat, http://www.mizzima.com/news/inside-burma/4-inside-burma/831-swarms-of-rats-destroy-crops-in-townships-in-chin-state.

Mowat, Farley (2010). "Never Cry Wolf Study Guide, 1963. Mizo Diaspora on Mautam Mission" *The Telegraph India*, June 13, 2008. http://www.telegraphindia. com/1080613/jsp/northeast/story_9402912.jsp.

Nashreen Mahbuba, 'Disaster Research: Exploring Sociological Approach to Disaster in Bangladesh'. [Online] in Bangladesh e-journal of Sociology Vol. 1 No. 2 July 2004: 1–8 [online] http://www.bangladeshsociology.org/Nasreen%20Sociology%20of%20Disaster, %20PDF.pdf [2 December 2005].

Oliver-Smith, Anthony and Hoffman, Susanna M. (1999). *The Angry Earth. Disaster in Anthropological Perspective*. New York/ London: Routledge.

Quammen, D. (1985). "Animals and Earthquakes: This World", *San Francisco Chronicle*, pp. 15–16.

Schaal, Rand B. (1988). "An Evaluation of the Animal Behavior Theory for Earthquake Prediction", *California Geology*, Vol. 41, No. 2, pp. 41–49.

Sen, A. (1981). *Poverty and Famines: An Essay on Entitlement and Deprivation*. Oxford University Press.

Shankar Darshan (2007). Personal Communication, Darshan Shankar, Foundation for Revitalisation of Local Health Traditions.

Siiger, Halfdan (1967). "The Lepchas, Culture and Religion of a Himalayan People, Part I, Results of Anthropological Field Work in Sikkim, Kalimpong, and Git", Publications of the National Museum. Ethnographical Series, XI. pp. 1–17.

Tamsang, K. P. (1998). "The Unknown and Untold Reality about the Lepchas", IInd edn. (Lyansang Tamsang), pp. 68–71.

The Telegraph India (2008)."Mizo Diaspora on Mautam Mission". June 13. http://www.telegraphindia.com/1080613/jsp/northeast/story_9402912.jsp.

United Nations Environment Programme (UNEP) (2006). "Knowing Nature: Indigenous Knowledge Helps in Disaster Management".

12

Natural Calamities and Native Perceptions

Evolving Effective Rehabilitation Strategies

Jesurathnam Devarapalli

Introduction

The human habitats are widespread across diverse environmental and physical contours and are supposed to have provided congenial space for comfortable living by making necessary adjustments with the help of sociocultural and technological apparatus. When the normal conditions get disrupted due to various extraneous forces so that social living becomes disturbed to a great extent and if the recouping ability of the people is inadequate to regain normalcy, such a situation have far-reaching consequences. This kind of disturbances are addressed through various terms such as disaster, hazard and calamity, which is a meeting point of nature, society and a human element in the form of their material and non-material product. In the words of Oliver-Smith and Hoffman (1999: 1) 'disaster is seen as a process leading to an event that involves a combination of a potentially destructive agent from the natural or technological sphere and a population in a socially produced condition of vulnerability'. The disasters thus project and promote the vulnerability of the inhabitants of that particular geographical area where the hazard occurred. It is important to realize the source of hazard, its magnitude and the impact it can inflict on the natural and social environment so to face it successfully. The responses to natural disasters may not be an all of a sudden reaction but it is a planned approach evolved out of accumulated exposures and experiences. The highest satisfaction of achievement is possible only if the victims are gratified in all the respects. This speaks of a state of resilience that reflects the ability of a community or society to accommodate and recover from the effects of a disaster as quickly as possible (Macalister 2015).

Disaster: A Perspective

Disasters are defined by the United Nations as 'a serious disruption of the functioning of a community or a society involving widespread human, material, economic, or environmental losses and impacts, which exceeds the ability of the community or society to cope with its own resources' (UNISDR 2009). The disasters are always challenging and invariably demands preparedness on the part of the victims (Kron 2000: 580). Post-disaster impacts are multiple and require specific strategies. Michael Lindell (2013) classified these adversities into physical and social impacts. Accordingly, the physical impacts include causalities and property destruction, while social damages account for psychological, demographic, economic and administrative implications.

Owing to the origins, disasters are naturally caused which may be further aggravated by anthropogenic activities such as deforestation, badly planned construction, developmental projects (Tyagi 2007), overexploitation of natural resources, various human endeavours that directly tilt the ecological balance, leading to climate change and so on. The natural disasters have their manifestations in various forms such as flood, drought, tsunami, earthquake, volcanoes, avalanches, epidemics and cyclones. The second category of disasters is directly linked to the human activities supported by technologies which are supposed to have commissioned to usher in human welfare but unfortunately lost control by its managers resulting in technological disasters. This kind of disasters includes nuclear fallout (Fukushima of Japan, Chernobyl of Russia), endosulfan spray in Kerala, industrial hazards (Union Carbide, Bhopal), use of sophisticated lethal weapon in warfares and many more such disasters. So, the calamities we confront with include both natural and technological disasters.

Brent Marshall and others (2004), quoting Barkun (1974) and Fritz (1961), distinguish the two by stating that natural disasters are 'acts of God' and technological disasters are 'human-caused'. This indicates that if human activities are properly channelized and applied with wisdom, technological hazards can be averted while the natural disasters though not stopped can be well mitigated and can minimize their ill-effects. The malfunction of a technological structure or human error in handling the technology or both ultimately results in technological disasters (Donovan et al. 2016), while natural disasters are based on natural phenomena or climatic factors.

Disasters are universal phenomena across the globe that inflict significant impact in varying degrees on both built environment and the natural environment. That is the reason why coping up with the disasters is not an easy routine, rather it involves much effort coupled with experience and expertise. Here the perceptions about the disaster makes all the difference in handling the situation. For the same or similar situation in different places or in the same place different responses are seen, owing to the cultural backdrop of the affected and/or the approach adopted by the mitigating agency. The reality of natural disaster exists in the images that people have of these events (Stallings 1988). How simple or small-scale societies respond to natural disasters are phenomenally different from those experienced by the technologically equipped societies. The former generally attribute these episodes to divine wrath and take the magico-religious routes to come out of the turmoil and/or sometimes try to move away to a considered safer location. While the latter deploy all-out effort using the technological know-how in dealing with the situation. In whatever be the social situation, satisfaction of the victims shall mark the success of relief and rehabilitation strategy.

Disasters and Anthropological Insights

Ever since ecological studies gained the momentum in anthropological research, apart from others, hazards and disasters have become the focus of analysis for various reasons such as academic, human welfare and testing hypothesis. Such studies have considered the natural disasters as sociocultural events rather than as just physical calamities. A good number of studies were carried out on droughts (Colson 1960; Lovejoy and Baier 1975), hurricanes (Firth 1959; Lessa 1964; Tonkinson 1979), floods (Hellman 2015; Zaman 1989) and frost (Margolis 1978; Waddel 1974). The approaches employed in disaster studies were broadly classified into two types according to Torry (1979). One is homeostatic approach which looks at the system regaining its normalcy through a battery of institutional safe guards. Second approach is developmental whereby adopting modernizing pressure that disrupt social stability and promote change.

Anthony Oliver-Smith (1996) in his review articie on hazards and disasters has pointed out that new perspectives were noticed in the recent writings which holds disasters 'as basic element of environment and as constructed features of human system rather than as extreme and unpredictable events as they were traditionally perceived' (p. 304). He further points out that 'recent perspectives in Anthropological research define a disaster as a process/event involving a combination of potential destructive agent(s) from natural and/or technological environment and a population in a socially and technologically produced condition of environmental vulnerability' (p. 305). According to him three perspectives were developed on disasters: (1) behavioural response, (2) social change and (3) political economic/environmental perspective. In spite of these approaches that can be realized from the existing literature, these studies also clearly focus on the pain, suffering, perception and expectations of the affected community which have a tremendous contribution to the evolving relief and rehabilitation strategies. The event of disaster not only brings in a breakdown in social and natural mechanism, but also leads to conflicts and confrontation owing to the perception held by the victims' social system. The view that conflict doesn't exist during disaster when compared to other situations is an opinion expressed by Quarantelli and Dynes (1976: 140). But this generalization according to Stallings (1988) is confined to a specific phase of emergency during the period of disaster but not necessarily to the earlier or later stages of disaster (p. 570). With this viewpoint he emphasized the conflict theory which holds that natural disasters are inherently contentious for the reason that political position always exert pressure on ruling dispensation for relief and rehabilitation. On the other side, the victims themselves insist on their likes and dislikes, throughout the course of relief and rehabilitation process.

Disaster Mitigation

Whenever a disaster occurs weather it is natural or technological, there should be adequate measures in place to cope up with the situation and to meet the grievances of the people. In providing such a succour government machinery shall take a statutory responsibility. But in real situations the private voluntary agencies of national or international origin plays a vital role in salvaging the victims from the calamity. However, in such circumstance the preparedness of victims of the hazard is of crucial importance (Kron 2000) in meeting the immediate needs and overcoming the initial damages and exigencies. In case of anticipated disasters due to electric short circuit, explosion, leakage in industrial units, disposal of biochemical wastes, installation of mega multipurpose projects and such others, an established crisis management procedures which are well known for the concern professionals are required to follow. Omissions and commissions regarding security protocol is an open invitation to disasters and such a behaviour needs stern action to compel preparedness. Even in case of natural disasters, some of them are recurring while some can be predicted, quite in advance. In either case, there is ample scope for the community on its own to prepare itself to meet the eventuality, thus minimizing the suffering. My observation on Ayodhya Lanka, a river island, in which the islanders are thoroughly prepared much in advance to face recurring floods. Before the onset of the monsoon, all the islanders irrespective of their socio-economic conditions acquire durable food materials, kerosene and firewood, and secure cattle to safe places in order to face recurring annual floods which may be normal or severe in nature. Not only that people are psychologically well prepared and do not express fear but they are also stubborn and unwilling even when the state administration is pressurizing them to vacate the island (Devarapalli 2008).

With the onset of disaster, the emergency starts which extend over some days or a few weeks of the post-disaster period. Depending on the nature of the disaster, the crises varies interrupting mundane activities of the people, dismantling habitats and even costing the lives of both human beings and animals. The normal life goes out of gear, the social networks may breakdown giving rise to complete dependency on the outside supporting systems. There may be much of pouring in of relief material in the form of food, water, medicine, blankets, utensils and other essentials. Ironically the relief can be alarmingly low and never reach in time or it may be immensely huge and got diverted. In each situation, politicians take political mileage and some voluntary agencies and government personnel involved in rehabilitation capitalize substantially (Das 2002). The post-disaster scenario often enhances the generosity of the co-citizens or people across the globe on learning about the misery of the affected. The deliverance of the relief is not properly monitored leading to the sufferings of some sections and overflow of materials for others. Due to the continuous supply of similar materials, some of the essentials may not be thought of by the providers, and hence the woes of the victims remain unresolved. In most cases, the entire relief material is the choice of the providers based on the generalized notion of the felt needs. But in reality, this may not be appealing. For example, all the agencies provide blankets or utensils but no one provides medicines or milk for immediate consumption. While some kind of materials distributed may not have utility value. During the emergency period of disaster, people are not only disoriented with their routine but also struck by mental agony and emotional problems that are not attended to nor recognized during the whole process of relief or rehabilitation that follows.

The needs of the people are based on their age, health, gender and other demographic characters, though we may ignore needs based on cultural factors in the initial period. The infants, crippled, sick and the aged are the worst victims because they are not capable of protecting themselves or make any effort to move away from the peril. Similarly, the women as in the words of Stewart Williams (2008) 'women especially suffered in the Tsunami, they have been further misunderstood, discriminated against, denied aid or sexually abused because of systematic inabilities to account for gender differences' (p. 1124). During the severe floods of Ayodhya Lanka, the women face the sensitive problems of attending nature's calls whether they are on the island or in the rehabilitation camps, which is an indication of non-relevance of gender throughout disaster and rehabilitation, which is a matter of no less significance. A higher fatality is noticed among children, elderly or sick following the 1970 tidal wave in Bangladesh and 2004 Tsunami in Asia (Ville de Goyet et.al 2006: 3). 'Stereotypes of gender vulnerability at the time of impact often do not apply. Depending on the type of the disaster, far more significant vulnerability factors than gender or age are the time of the day of impact ... including the location of the victims within the buildings' (Ville de Goyet et al. 2006: 3). It is evident that many such sensitivities significant at personal level and well taken care of in normal situations would become null and void as safeguarding life became more important than other things. As the severity of the disaster declines and the threat for life is averted then comes to the surface umpteen number of issues which require a decisive strategy to rectify damage caused to the people and their properties, and put the social and economic system back on the track.

Rehabilitation and Reconstruction (RR)

It is a phase in the disaster that the administrative machinery, voluntary agencies, professionals, planners and a host of others come together and surround the victims with a single goal to overcome all the hurdles created by the disaster and provide them with the pre-disaster facilities or something better than that. Everyone becomes vigorously engaged in their specialized activity and trying to give the best out of their expertise to the victims on one side. But on the other hand, there are efforts to make maximum unlawful encashment at the cost of the project which may end up in injustice to the target group. The post-disaster efforts involve both temporary and permanent measures. The first category involves providing medicines, disposing the debris, removal of carcasses or biological wastes, and restoration of water, electricity, communications and other amenities that are very essential in sustaining lives and keep them fit. While the second category contains reconstruction of the houses or dwellings, resettlement/displacement, creation of livelihoods, introduction of new institutions for a better life and so on; these have a long-lasting effect on the victims themselves and on their culture. In giving the instantaneous help or in clamping the long enduring strategies, acknowledging and accommodating the victims' likes and dislikes should be a hallmark. Otherwise much of effort and materials offered would go unavailed by the target groups. In addition to that, practically, when no opinion or view of the victims are neither tapped nor incorporated in the RR programme it ultimately distances the victims from the restoration activities. In order to illustrate this argument some of the earlier works are referred that clearly portray the people's resentment and dissatisfaction towards the kind of support created in their crisis state.

Natural disasters are such social events where structural, functional and organizational features of social system are subjected to critical experimentation that would demonstrate their abstract existence and dispel the illusions surrounding societal controls over the individual. Disasters not only destroy the natural habitat but also tear the social fabric by breaking the families with sudden death of its members and eliminating many persons who are expected to discharge certain roles in the society to keep up the normal course of life in motion. In spite of the loss of human life, economic resources and working abilities, the society still tries to cope up with the situation. But, however, the fissures in the structural and organizational frame are visible as surviving people find it difficult to imitate the original pre-disaster state of stability. Under these conditions, invariably the external assistance is desired by the victims but cannot be at the cost of their aspirations and expectations. The outsiders' perception of needs of the affected is different from that of the victims. In that sense the views from the community are expected for an effective implementation of aid distribution. But when the locals are involved it is generally the elite or the leaders of the group that dominate the entire decision making and even diverting the aide to the benefit of their group members or their kin and close followers, thus rendering disservice to the larger section of the society (Leemann 2013). My study on the flood situation of Ayodhya Lanka clearly shows that the local leaders involved in assisting the administrative machinery in enumerating the damage always favour their own caste men to get undue benefit irrespective of the actual damage they sustained. Even fully damaged houses of other caste groups especially Dalits were paid little attention if they are not their loyalists. A free hand given to the community leaders would end up in utter irregularities in relief distribution, though they are shrewd enough to project concerns and requirements of the people. The personnel of the local government should be very vigilante and judicious enough to check the biased reporting of the community leaders and try to directly witness and hear from the sufferers. Also, the commitment and uprightness of the officer who are entrusted with their task matters supreme in all this exercise.

Often it is a common scenario that whenever a disaster occurs the local or the outside NGOs plunge into action uninvited. They are supposed to render selfless and impartial service to the people owing to the ideal commitment they have taken. However, there are contradictory views, as pointed out by Haberli (2013: 38), that membership with the NGO becomes essential for the victims' names to be entered into the damage and needs list in order to access the external aid. But there are also quite a good number of references that put on record the laudable efforts of NGOs in getting adequate help to the satisfaction of the recipients (Anderson and Woodrow 1989; Mantel

2013). The people, though under the pressures of emergency in that critical time, agree for any proposal that comes but in the course of time they realize the inconsistencies and impediments with the entire package. There is a lot of wastage with regard to the relief provided spontaneously for the immediate use and consumption, which is never visible or accounted for. But this is clearly visible in the case of long-lasting relief arrangements such as resettlement, livelihoods and house type and the resultant agony caused is more grievous than the disaster itself.

Destruction of shelter is a major consequence due to the natural disasters, and it is not only the concern of the victims but it also draws the attention of various agencies such as NGOs, government and relief funding bodies. These agencies either independently or in collaboration with one another comes up with a blueprint of the dwellings as they have generated resources for the purpose. The quantum of funding and the house type and its design goes in tandem. But what is significantly missing is peoples' likes and expectations. People generally shaped dwelling in tune with their living conditions, work environment, maintenance capability and their world view. Irrespective of the funding sources whosoever is facilitating the house construction either it may be contractor or NGO, unless it is owner-driven reconstruction, the highest levels of satisfaction and optimum utilization of the structure is not possible (Duyne Barenstein 2013: 75). The active involvement of the owner in the house construction is also beneficial because work quality is monitored, could meet time line and reflect at least some of their desires on the super structure provided the builder is graciously flexible to the owner. Though the participation of the owner in reconstruction gives a moral boost, but practically the impact may not be that effective (Sliwinski 2010) as their views and suggestions are relegated and ignored. In the flood-prone Ayodhya Lanka, houses are built on elevated platforms artificially created by dumping soil to a height of 5–6 feet, so that the houses are not easily inundated unless the flood water raise to that height. The houses are built with locally available materials such as palmyra leaves, grass and tree branches and of course bamboos from outside. All the houses are built with these materials irrespective of the socio-economic condition of the owner. The reason is that the damages due to recurring floods to these houses would be minimal and its reconstruction after damages is affordable to a large extent by the inhabiting family members. So, when the issue of finding a permanent solution is considered, the islanders were dead against vacating the island and opt for resettlement on the main land. Construction of independent modern houses with measures to tide over inundation is very expensive. Therefore, the planner adopted the local housing design and built two shelters across the island as rescue centres during floods and cyclones. These are concrete structures with a cellar over which at a height of 7 feet huge hall is built for people to take shelter during inundation and the cattle can be rescued in the cellar which is for all time use.

The natural disasters would influence the government decisions in uprooting the dwellings and resettling them in a new place which was supposed in their view to be safe and guarded from the recurrence of such disaster. But this idea of replacement, in many ways, displeases the victims for various reasons such as separation from their natural habitat, alienation from their traditional occupation, sociocultural incompatibility in the new habitat, scarcity of livelihoods, vulnerability and marginalization (Ganguly 2005). As in the case of Gujarat earthquake of 2001, out of the 243 villages reconstructed in Kutch by NGOs only 23 were relocated and of these only one village relocation has justification that it was inundated due to the proximity of dam (Duyne Barenstein 2013). The earthquake victims of Gujarat vehemently opposed the resettlement proposal forcing the government to drop that idea, and many of the houses that were reconstructed were not occupied by the people as they suspect the quality of construction and the layout do not suite their rural lifestyle, also there is lack of privacy for women, lack of space for cattle and no social ambience (Duyne Barenstein 2013). The *pucca* houses constructed due to 2004 tsunami, for fisherman of Kalapet in Puducherry, are not occupied as they are located about 3 km from the beach where their economic activity operates and hence they remain in the beach-based old settlement as earlier. The emotional attachment to the old habitat, the natural landmarks such as water bodies, special type of useful vegetation and the socio-economic importance of these resources are the strong bonds of attachment. To this effect the study of Naimi-Gasser (2013) in Tamil Nadu where 5000 trees of mango, tamarind, coconut and others were removed in the process of reconstruction after tsunami is a very good anecdote. She pointed out that due to the change in physical environment which in turn affected the social life of the people resulted in psychosomatic disorders and depression especially among women. This kind of dislocation from their original habitat and/or destruction of the habitat would have strong negative impact on the economic condition of the people. The new type of economic activities offered by the NGO or rehabilitating agencies need not be readily accepted by the people for the lack of competence to handle the new enterprise or it may not be having long-lasting gains and the people have to struggle for the alternative means. Many times, the facilities created in the RR package may not be very familiar to the local people and not readily accepted, and the facility would go waste. Especially in the Indian scenario when modern toilets were introduced and wash basins were created, people could not use them rather they misused these and ultimately, these were rendered useless. Any damage caused to the house could not be addressed by the owner because it is not in their skill, and the building material is also alien to them.

The RR package not only transformed the physiography of the traditional villages but have also dented on the ownership rights of the newly built houses. It may sometime happen that marriage is treated as criteria for allotment of new house. But a nuclear family with four grown-up sons who were not married though had reached the marriageable age gets only one allotment which is so small to accommodate all the six adults, and practically there is no individual privacy. This at times would prompt the victims to provide false and fictive information or claim for the allotment through dubious marriage certification. Another significant situation with house allotment is that the elderly, divorced and singulars are not considered expecting that their children or primary kin would take care of them. But the authorities concern fail to understand that without an asset no one would entertain these elderly, thus rendering them marginalized and disempowered (Duyne Barenstein and Trachsel 2013: 160).

Conclusion

Disasters, impertinent of their origin either due to natural/climatological or technological causes, are a test of time for sociocultural resilience of any human population. This brings in together three apparently different elements, namely, nature (or cause of the disaster), human populations (victims) and means of relief (technology, culture, etc.) which are very much inter influencing irrespective of the type of disaster and its frequency, disasters always disturb normal course of human life which may have to be repaired or rebuilt. A number of agencies are spontaneously created or pre-existing plunge into action to provide relief so to restore normal situation before the people go into undue sufferings. There is local level, national level and international bodies to mitigate disaster situation. These agencies emerge and operate either under the statutory acts of the concern nation/government or United Nations organization, with well-planned strategies, commitment and expertise. Once the emergency phase is over whatever is left behind by the disaster has to be well secured and restored so that the remaining people and their habitat are assured of returning to pre-disaster reality. With so much of funding in the form of money and material at hand, the agencies are in brisk action to create the felt needs to the prevailing situation. The immediate relief is sought for food, water, clothes, medicine and shelter to meet the emergency that prevails over the initial few days of the disaster. Coming to the long-lasting measures unlike the short term, they cannot be decided and imposed by the relief providers. Particularly with housing, resettlement and livelihoods, a greater role should be given to the victims to decide upon and choose what they want. Otherwise what has been created against the taste may go waste as the people find the given facility is in noncompliance with their working and living conditions or they found themselves lacking in managing the given facility. Though the relief agencies accept the idea that they should involve the victims in the process of designing the RR package but outcome is not reflective of the people's views. This clearly shows that the involvement of the victims in the RR activities is only titular but it doesn't cause any weightage in the course of execution of the entire project. Rather the participatory approach which is meant for reflecting needs and aspirations of the targeted people is in fact used as a tool for giving higher legitimacy to the project that by giving the impression that the locals were involved throughout the course (Duyne Barenstein and Leemann 2013: XXVIII). When the people's views were not reflective they feel let down, though whatever the best is provided to them. Contrary, however conservative the people are, the acceptance or rejection of the new innovation depends on how best the people are informed of it. Further, it is a common experience that the people often revert back to their traditional ways though unsustainable, if timely inputs are not given and the system that ensure the continuity of material and skill are not in place in right time (Niazi 2016). Throughout the course of post-disaster relief and restoration strategies, the demographic criteria and the gender-linked cultural features of the victims should be invariably given due weightage so that the entire effort put in will be materialized when it is welcomed and accepted by the target group.

References

Anderson, M. and. Woodrow, P. (1989). *Rising from the Ashes: Development Strategies in Times of Disaster*. Boulder CO: Westview Press.

Barkun, M. (1974). *Disaster and the Millennium*. New Haven, CT: Yale University Press.

Colson, E. (1960). *Social Organization of the Gwembe Tonga*. Manchester: Manchester University Press.

Das, K. (2002). "Social Mobilization for Rehabilitation: Relief Work in Cyclone Affected Orissa", *Economic and Political weekly*, Vol. 37, No. 48 (November 30–December 6), pp. 4784–4788.

Devarapalli, J. (2008). "Natural Disturbances and Cultural Response on a River Island in Andhra Pradesh", *Journal of Ecological Anthropology*, Vol. 12, No. 1, pp. 5–21.

Donovan, M., Smith, S., Radanovich, H., and Gutter, M. (2006). "Impact of Technological Disasters". Available at https://edis.ifas.ufl.edu/pdffiles/FY/FY123000.pdf (Accessed on December 20, 2016).

Duyne Barenstein, J. (2013). "Communities' Perspectives on Housing Reconstruction in Gujarat Following the Earthquake of 2001", in *Post-Disaster Reconstruction and Change: Communities' Perspectives*, ed. J. Duyne Barenstein and E. Leemann. Boca Raton/London: CRC Press, Taylor and Francis Group.

Duyne Barenstein, J., Jennifer, E., and Leemann, E. (eds.) (2013). *Post-Disaster Reconstruction and Change: Communities' Perspectives*. Boca Raton/London: CRC Press, Taylor and Francis Group.

Duyne Barenstein J. and Trachsel, Sonja (2013). "The Role of the Informal Governance Post-Disaster Reconstruction and Its Impact on Elderly People's Social Security in Coastal Tamil Nadu", in *Post-Disaster, Reconstruction and Change: Communities' Perspective*, ed. J. Duyne Barenstein and E. Leemann. Boca Raton/London, CRC Press, Taylor and Francis Group.

Firth, R. (1959). *Social Change in Tikopia*. New York: Macmillan.

Fritz, C. E. (1961). "Disaster", in *Contemporary Social Problems*, ed. R. K. Merton and R. A. Nisbet. New York: Harford.

Ganguly, V. (2005). "Displacement, Rehabilitation and Resettlement: The Case of Maldhari Families of Gir Forest", *Sociological Bulletin*, Vol. 54, No. 1 (January–April), pp. 3–17.

Haberli, I. (2013). "Aid Distribution after Hurricane Mitch and Changes in Social Capital in Two Nicaraguan Rural Communities", in *Post-Disaster, Reconstruction and Change: Communities' Perspective*, ed. J. Duyne Barenstein And E. Leemann. Boca Raton/London: CRC Press, Taylor and Francis Group.

Hellman, J. (2015). "Living with Floods and Coping with Vulnerability", *Disaster Prevention and Management: An International Journal*, Vol. 24, No. 4, pp. 468–483.

Kron, W. (October 2000). "Natural Disasters: Lessons from the Past—Concerns for the Future. The Geneva Papers on Risk and Insurance", *Issues and Practices,* Vol. 25, No. 4, pp. 570–581.

Leemann, E. (2013). "Communal Leadership in Post-Mitch Housing Reconstruction in Nicaragua", in *Post-Disaster, Reconstruction and Change: Communities' Perspective*, ed. J. Duyne Barenstein and E. Leemann. Boca Raton/London: CRC Press, Taylor and Francis Group.

Lessa, W. (1964). "The Social Effects of Typhoon Ophelia (1960) on Ulithi", *Micronesia*, Vol. 1, pp. 1–47.

Lindell, M. K. (2013). "Recovery and Reconstruction after Disaster", in *Encyclopedia of Natural Hazards*, ed. P. T. Bolrowsky. pp. 812–824.

Lovejoy, P. E. and Baier, S. (1975). "The Desert Side Economy of the Central Sudan", in *The Politics of Natural Disaster: A Case of the Sahel Drought*, ed. M. H. Glantz. New York: Praeger. pp. 145–175.

Macalister, F. (2015). "Preparing for the Future: Mitigating the Disaster and Building the Resilience in the Cultural Heritage Sector", *Journal of the Institute of Conservation*, Vol. 38, No. 2 (October), pp. 115–129.

Mantel, C. (2013). "Ownership Control and Accountability in Post-Tsunami Housing Reconstructive Process is Aceh, Indonesia", in *Post-Disaster, Reconstruction and Change: Communities' Perspective*, ed. J. Duyne Barenstein and E. Leemann. Boca Raton/ London, CRC Press, Taylor and Francis Group.

Margolis, M. (1978). *The Morning Frontier: Social and Economic Change in a Southern Brazilian Community*. Granville: University of Florida Press.

Marshall, B. K., Steven, P. J., and Schlichtmann, J. R. (April 2004). "Technological Disasters, Litigation Stress, and the Use of Alternative Dispute Solution Mechanisms", *Law and Polity*, Vol. 26, No. 2, pp. 289–307.

Naimi-Gasser, J. (2013). "The Remembered Trees: Contractor-Driven Reconstruction and Its Consequences on Communities' Well-being in Coastal Tamil Nadu", in *Post-Disaster Reconstruction and Change: Communities' Perspectives*, ed. J. Duyne Barenstein and E. Leemann. Boca Raton/London: CRC Press, Taylor and Francis Group.

Niazi, Z. (2006) "Reconstruction and Rehabilitation: A Response Strategy for Creation of Sustainable Livelihoods". Available at http://www.devalt.org/newsletter/may01/lead.htm (Accessed on December 20, 2016).

Oliver-Smith, A. (1996). "Anthropological Research on Hazards and Disasters", *Annual Review of Anthropology*, Vol. 25, pp. 303–328.

Oliver-Smith, A. and Hoffman, S. M. (eds.) (1999). *The Angry Earth: Disaster in Anthropological Perspective*. New York: Routledge.

Quarantelli, E. L. and Dynes, R. R. (1976). "Community Conflict: Its Absence and Its Presence in Natural Disaster", *Mass Emergencies*, Vol. 1, pp. 138–152.

Sliwinski, A. (2010). "The Politics of Participation: Involving Communities in Post-Disaster Reconstruction", in *Rebuilding After Disaster: From Emergency to Sustainability*, ed. G. Lizarralde, C. Johnson, and C. Davidson. London: Spon Press.

Stallings, R. A. (1988). "Conflicts in Natural Disasters: A Codification of Consensus and Conflict Theories", *Social Science Quarterly*, Vol. 69, No. 3 (September), pp. 569–586.

Tonkinson, R. (1979). "The Paradox of Permanency in a Resettled New Hebridean Community", *Mass Emergencies*, Vol. 4, pp. 105–116.

Torry, W. L. (1979). "Anthropological Studies in Hazardous Environment: Past Trends and New Horizons", *Current Anthropology*, Vol. 20, No. 3, pp. 517–540.

Torry, W. L. (2001). "Natural Disasters, Social Structure and Change in Traditional Societies", *Journal of Asian African Studies*, Vol. XIII, pp. 3–4.

Tyagi, A. (2007). "Global and National Scenario of Natural Disasters: With Special Emphasis on Floods", *The Indian Journal of Political Science*, Vol. 68, No. 4 (October–December), pp. 791–808.

UNISDR (United Nations International Strategy for Disaster Risk Reduction). (2009). *UNISDR Terminology on Disaster Reduction*. Geneva: UNISDR (available online)

Ville de Goyet, C., Marti, R. Z., and Cesorio, Claudio (2006). "Natural Disaster Mitigation and Relief", in *Disaster Control Priorities in Developing Countries*, ed. D. T. Jamison, J. G. Breman, A. R. Measham, et al. Washington, DC: The World Bank and co-published by Oxford University Press.

Waddel, E. (1974). "Frost Over Niugini: A Retrospect on Bungled Relief", *New Guinea,* Vol. 8, pp. 39–49.

Williams, S. (2008). "Rethinking the Nature of Disaster: From Failed Instruments of Learning to a Post-Social Understanding", *Social Forces*, Vol. 87, No. 2, pp. 1115–1138.

Zaman, M. Q. (1989). "The Social and Political Context of Adjustment of River Bank Erosion Hazard and Population Resettlement in Bangladesh", *Human Organization,* Vol. 48, pp. 196–204.

13

Disaster Diplomacy in South Asia

India's Humanitarian Assistance Programmes in Neighbourhood

Sibaram Badatya

Introduction

Among numerous challenges that human civilization face today, natural disaster remains the most destructive force and causes more fatality than any other threat. According to Centre for Research on the Epidemiology of Disasters (CRED) database, about 22,455 natural disasters occurred since the beginning of twentieth century killing more than 38.57 million people and injuring another 8.7 million people. In total, more than 7.5 billion people were affected and the property of nearly USD 3 billion was destroyed between 1900 and 2016 due to the nature's fury (CRED 2017). The Asian continent, being the worst disaster-prone region (Bhatiya 2014), witnessed approximately 9,351 disasters affecting nearly 6.7 billion people of which, about 27 million lost their life and injured another 5 million. The total damage of the property is also high as compared to any other continent, pushing the region to further backward. Among regions of Asian continent, South Asia remains the most disaster-prone area.

Constricts of seven nations—Bangladesh, Bhutan, India, Maldives, Nepal, Pakistan and Sri Lanka—South Asia is the one of the poorest region of the world and the home for one-fifth of world's population (Gilbert 2017). Along with high pollution, this region is also sensitive to natural disasters, witnessing more than 900 events since 1970 alone.[1] People of this region has been victimized by periodical disasters including cyclones, droughts, earthquakes, floods, landslides and tsunamis. In recent times, this region has witnessed numerous major natural disasters, including 1999 Odisha super cyclone, 2001 Gujarat earthquake, 2004 Indian Ocean tsunami, 2005 Kashmir earthquake, 2007 Bangladesh cyclone, 2008 Baluchistan earthquake, 2010 Pakistan floods, 2013 Uttarakhand flood, 2014 Kashmir floods and 2015 Nepal earthquake.

In the subcontinent, the menace of natural disaster affected more than 750 million between 1990 and 2008 resulting in almost 230,000 deaths and about damaging property of US $45 billion. Flood, which counted for half of the total number of hazards, caused severe economic damages to the region, while droughts accounted for 2 per cent but that affected more than 50 per cent of the total number of affected victims.[2] Due to the vulnerable geographical landscape, comprising the high Himalayas at the north and surrounded by seas at south, South Asian region makes it particularly susceptible to natural disasters. All the countries of South Asia are marked as the high disaster-prone states and continue to occupy high ranks in the World Risk Index. In the recently published 2016 World Risk Report, for instance, Bangladesh ranked as the fifth most risk-prone country of the world, while Afghanistan, Bhutan, Sri Lanka, Pakistan and India ranked 41, 60, 63, 72, and 77 respectively.[3] There is a high level of variation within South Asia itself, but these rankings should not be misinterpreted to undermine the risk faced by these countries—particularly the poor—in the face of natural disaster.

Besides the human losses, the menace of disaster also contributed to the prolonged poverty and poor living condition in the vulnerable regions of the world. As the people of developing countries lacks standardized housing and anti-disaster infrastructures the destruction is much higher than the developed countries pushing the countries further back. In addition, the poor infrastructure and high human settlement in the low-lying areas make the region more vulnerable to natural disaster. Moreover, the wide economic disparity and inadequate governmental support left this region vulnerable to natural disasters. Most of the countries of the subcontinent lack effective relief mechanism, emergency medical facility and subject to widespread corruption, misgovernance for decades. As result, South Asian states ranked high in terms of disaster coping and adaptive capacity. Afghanistan, for instance, has been ranked number one country in lacking disaster coping capacity and fifth in terms of adaptive skills in the world, making it the fourth country in world in overall disaster vulnerability. Similarly, Pakistan with 62.48 points occupies fourteenth ranks in lacking adaptive capacity.[4] The position of other South Asian countries is encouraging. While Bhutan and Sri Lanka occupies 60 and 63 in overall Disaster Risk Index respectively, both the countries also score high (73.77% and 78.08% respectively) in lacking adaptive capacity.[5] India and Nepal, which got relatively good ranks (77 and 108) in terms of overall Disaster Risk Index case also score poor (80.22% and 81.05%) in terms of lacking adaptive capacity. This poor state of disaster lacking capacity is largely related to weak economic status of the region. However, in some case, it is because of the respective state government and their prolonged negligence in strengthening the disaster adaptive capacity of a state despite impressive economic growth, such as in India. As a result, the rate of casualty remains high in South Asian region.

India's Disaster Diplomacy in South Asia

Being the largest state in the South Asian region, India faced menace of disaster periodically. Despite a Third World developing country, India is able to prevent the destruction of natural disaster with its limited resources and has shown a successful model for South Asian countries and beyond. Apart from the sphere of influence, New Delhi also extended its disaster expertise and sent its men and machineries into the ground of the neighbouring states to assist the local administration in disaster response operations. From Afghanistan to Myanmar and from Bhutan to Maldives, India has been actively participated in various humanitarian assistance programmes in recent past. Though India's disaster efforts during recent Nepal earthquake got the world attention, New Delhi has been engaged in trans-national relief operations in the past. India's offer of help during Pakistan earthquake in 2005, its assistance during 2004 tsunami in Sri Lanka, Bangladesh and Maldives, and various relief efforts in monsoon flood and deadly cyclones in Bangladesh, Sri Lanka and Myanmar have earned much goodwill for India and enhanced its soft power image in the neighbourhood. However, the magnitude of India's relief programme was low due to its economic and technical resources. However, it has today shown the world that it can handle crisis like the one she faced in Yemen in evacuating its beleaguered citizen from an active war zone and in Nepal where massive humanitarian assistance was carried out by India. Hence, it is essential to discuss, in detail, India's disaster assistance programme in the neighbourhood.

The biggest ever relief operation called 'Operation Maitri' was launched and the recent rescue operation in Nepal extended both financial and technical assistance at the time of human suffering. Besides 1 billion aid assistance, India has mounted a mammoth rescue and relief effort deploying its sophisticated men and machineries including National Disaster Relief Force (NDRF) personnel, medical team, and a number of aircrafts, helicopters, and civilian and military vehicles. The Indian Air Force has airlifted mobile hospitals, engineer task force with special equipment, specialized teams of NDRF along with tons of relief material including blankets, tents, tonnes of food, paramedics, stretches and medicines.

Indian Disaster Assistance Programme in Nepal

Nepal has been continuously affected by the menace of natural disaster for decades. In recent years, the frequency of natural disaster increased drastically. For instance, two deadly floods occurred in the year 2007 and 2008, affecting about 439,952 and 180,000 people respectively. Similarly, the landslide in Seti River resulted another severe flood in Nepal in May 2012 killing at least 15, and another 36 more missing. However, the biggest ever natural disaster occurred in Nepal was on 25 April 2015, when a devastated earthquake of 7.9 magnitude shock the Himalayan republic killing at least 7,365 people and injuring 14,355 others. It is the most powerful disaster to strike Nepal since the 1934 Nepal–Bihar earthquake. More than 8 million people were affected by the earthquake—roughly a quarter of the country's population—the United Nations said. According to one charity, UNICEF, 1 million of those affected are children. Being a small, poor and landlocked country between India and China, Nepal requested international help at this great tragedy.

India, being its immediate neighbour, offered all kinds of help during the national disaster of Nepal. India was the first responder to the crisis and started the relief and rescue operation within hours of the quake by involving Nepali ex-servicemen from India's Gurkha

Regiments as interface for guidance, relief and rescue. Prime Minister Narendra Modi directed immediate dispatch of relief and rescue teams, including medical team, to Nepal. Hence, by the afternoon of 25 April, 10 teams from India's NDRF, totalling 450 personnel, with search and rescue dogs reached Nepal and stated 'Operation Maitri'. Next day, six more NDRF teams along with several medical teams joined the rescue operation. In addition to the NDRF personnel, India has mounted a mammoth relief effort deploying 13 military aircrafts including C-17 Globe Master, C-130 Hercules, IL-76 and three civilian aircrafts of Air India and Jet Airways, six Mi-17 helicopters, two Advanced Light Helicopters. Indian Air Force also airlifted mobile hospitals, engineer task force with special equipment, specialized teams of National Disaster Relief Force (Singh 2015). By 26 April, the Indian Air Force evacuated more than 500 citizens from Nepal. During the rescue operations they have not just rescued Indians, but citizens of other countries as well. In addition, Air India reduced fares on flights bound for Kathmandu from Delhi, Kolkata and Varanasi, and announced that it would also carry relief material on its flights.

In addition to the relief materials, India sent 43 tons of relief material, including tents and food within the immediate aftermath of the quake. By 26 April, India dispatched a further 10 tons of blankets, 50 tons of water, 22 tons of food items and 2 tons of medicines to Kathmandu. Nearly 1,000 NDRF personnel were also pressed into service, and a 'big evacuation' of Indian citizens was underway through the road route. The government deployed 35 buses to evacuate stranded Indians in Nepal by road along the Indo-Nepal border. India started issuing goodwill visas to foreigners stranded in Nepal and mobilizing buses and ambulances to bring them to India by road Indian Railways provided 100,000 bottles of drinking water to be delivered by the Air Force, with arrangements being made to supply 100,000 bottles every day (*DNA*, 26 Apr 2015).

By the morning of 27 April, the Indian Air Force had evacuated 1,935 Indian citizens from Nepal using 12 aircraft sorties. The Indian army successfully evacuated 170 foreign nationals from the United States, the United Kingdom, Russia and Germany. Indian Army sources said 6 of 18 medical teams, comprising 20 people each, tasked to help with the relief efforts in Nepal had been deployed. The Indian Army was in the process of sending 10 engineer task forces of about 60 people with machinery to clear roads and debris. The troops have taken with them a further 10,000 blankets and 1,000 tents were on standby. The Indian army was also bringing oxygen cylinders for distribution to medical teams. Spain sought India's help in evacuating its nationals from Nepal. Prime Minister Narendra Modi promised help (Petroff and Rooney 2015).

Following the improvement of weather by 28 April, the Indian Air Force and the Indian Army fanned out to remote areas while continuing to transport essential items from various bases to Kathmandu and evacuating distressed persons to India. A 41-member medical team, along with medical supplies, was sent to Nepal from Rajasthan. The Sashastra Seema Bal dispatched more than three dozen vehicles, including ambulances and water tankers to Nepal from its border camps. The Indian government dispatched a further 220 tons of food packets and dry rations, 50 tons of water, 2 tons of medicines, 40 tents and 1,400 blankets to Nepal (Petroff and Rooney 2015).

Besides the relief and rescue operations, India also pledges USD 1 billion aid for Nepal's reconstruction programme, which is the highest among all donors. China, with much bigger economy, has only provided 20 million yuan (USD 3.3 million) in the initial period and pledged USD 9.9 million in total (*Xinhua News* 26 April 2015). The United Kingdom, who have a special relationship and proudly poses Nepali solders as part of Gorkha regiment, contributed only $130 million (£83 million), of which $51 million (£33 million) was donated by the government and $79 million (£50 million) was donated by the public. World's biggest economy United States only offered $10 million as a humanitarian assistance. Apart from the unparalleled generosity, Prime Minister Narendra Modi personally oversaw the whole operations and took all necessary steps during the great sufferings of Nepalese people. India's post-disaster efforts in Nepal were appreciated by world communities and generated goodwill between New Delhi and Kathmandu.

India's Humanitarian Assistance Programme in Sri Lanka

India and Sri Lanka share a very warm and friendly relation since ancient time. Being the only northern and the biggest neighbour of the island nation, India always stand with Sri Lanka in the difficult moments. India actively participated in the humanitarian and developmental activities of Sri Lanka. For instance, during the early years of civil war, India supplied food and medicine in Jaffna regions of Sri

Lanka by air and sea. India also actively participated in the negation and reconciliation process between Sri Lankan government and the LTTE. After subsequent negotiations, India and Sri Lanka signed an agreement on 29 July 1987, which ensures a certain degree of regional autonomy in the Tamil areas with right to self-government by their own regional council. By this peace accord, India also promises for the end of insurgency by perusing the groups to lay down their arms and sent the Indian Peacekeeping Force (IPKF) to Sri Lanka to enforce the disarmament and to watch over the regional council. As promise, the Indian peace keeping force reach Sri Lanka to maintain peace and for the smooth conduct of provincial election. However, the rejection of peace proposal led to the war between IPKF and LTTE resulting in heavy loss of Indian forces. The 32-month presence of the IPKF in Sri Lanka resulted in the deaths of 1,200 Indian soldiers and more than 5,000 Sri Lankans. The cost for the Indian government was estimated to be approximately 10.3 billion (Chattopadhyaya 1994; Richardson 2005). Since the withdrawal of IPKF, India kept itself away from all the internal affairs of Sri Lanka and its developmental activities in the island nation was limited.

However, the Indian Ocean tsunami of 26 December 2004, which killed about 227,898 people in total and injured about 125,000, besides displacing approximately 1.74 million people of the Indian ocean region, gave a new opportunity to India to step up its humanitarian assistance programme in Sri Lanka. During the worst ever tsunami, approximately 30,196 people of Sri Lanka alone died and about 21,411 people injured leaving approximately 516,150 people displaced (BBC 3 January 2005). The total destruction of property in Sri Lanka alone totalled about $1.30 billion (*The Guardian*, 25 December 2014). Following the tsunami, India officered all kinds of human assistance to Sri Lanka and pledged USD 23 million dollars as disaster relief aid (Lancaster, 2005). In addition to the monetary assistance, India also step up its rescue operation, and Indian naval helicopters reached Sri Lanka with immediate relief materials within 12 hours of the tsunami. By 27 December, two Indian Naval ships, INS Sharada and INS Sandhayak with full of relief materials, reached Galle and Trincomalee respectively, followed by INS Sutlej on 28 December. Indian navy medical team arrived in Colombo within hours of the tsunami and set up several medical camps for the disaster victims. Within weeks of the tsunami, India has deployed 14 ships, nearly 1,000 military personnel and several dozen helicopters and airplanes to its devastated island neighbour. India also sent a team of Indian divers to Sri Lanka for raising the sunken vessels that were preventing relief ships from entering the port, which have been completed in 8 days with the help from the Sri Lankan navy. The army engineering unit of Indian army completed a number of projects in the area, including restoring electricity to several government buildings that were inundated by the ocean surges. In addition, it has pledged $23 million for post-tsunami reconstruction projects in Sri Lanka (Lancaster 2005).

Although coordination of relief efforts was subsequently ceded to the United Nations, India and its partners in the aid operation have continued to coordinate closely. Berry, the Indian defence attaché to Colombo, held daily meetings with all the state holders of Sri Lankan government and played a crucial role in the rescue and relief operation. Indian relief operation in Sri Lanka was organized and effective that Bonhomme Richard, a Navy helicopter carrier, which was scheduled to assist the Sri Lankan relief activities, decided to changed course for Indonesia, witnessing India's early assistance. As per the Washington Post, additional US troops were no longer needed in Sri Lanka because the Indian military had been dispatched to help.[6]

The end of civil war after the defeat of LTTE once again opened the door for Indian assistance programme in Sri Lanka. India actively participated in the post-war developmental activities in war-torn regions and pledged ₹ 8 billion towards the reconstruction projects in the war-torn regions of Sri Lanka. India is active in a number of developmental projects in Sri Lanka which amount to one-sixth of India's total international development credit. After the civil war, India actively engaged in the rehabilitation and resettlement of the Internally Displaced Persons (IDPs). It supplied 250,000 livelihood packs to IDPs comprising food, clothes and medicine as immediate assistance after the war in May 2009. For the safe return of the displaced people, India deputed seven demining teams to clear the mines that have been laid down in the war-turned northern regions of Sri Lanka during the decades of civil war. With the help of local deminers, the Indian team worked for 3 years to clear large areas of northern area. The project cost about LKR 1 billion (INR 40 crore) and facilitated the resettlement of 300,000 IDPs.[7] As an effort to speed their early settlement, India provided about 10,400 MT of GI sheets of approximately 1.50 million GI sheets with the cost of LKR

1.3 billion (INR 50 crore) and 400,000 cement bags (20,000 metric tons) of LKR 199 million (INR 8.3 crore) which help approximately 80,000 and 49,000 displaced families to construct temporary shelters immediately. Each family was supplied with 8 cement bags and 12 corrugated sheets which enabled them to construct temporary homes on their own (High Commission of India in Sri Lanka n.d.).[8]

Based on the request from the Sri Lankan Government, India renovated of 27 schools of the Northern Province of Sri Lanka with the estimated cost of SLR 250 million (INR 12 crore) and constructed a Skill Development Centre and Civil and Mechanical Engineering Campus at University of Jaffna with the project cost of SLR 300 million (INR 14.3 crore) and SLR 256,028,440/- (INR 14.3 crore) respectively to empower the people of Sri Lankan people. In order to quickly revive farming activities in the north, India adopted a project of SLR 224 million (INR 14 crore) and supplied agricultural implements and seeds to farmers. It also provided tractors to agro-centres across the five districts of the Northern Province. This project helped approximately 95,000 farming families and strengthened their livelihood in the war-turned island nation. Indian government contributed seven demining teams to facilitate the movement of the people and their early resettlement in the region. To supplement this anti-mine action, artificial limb-fitment camps were organized to address the needs of the victims of mines and other unexploded ordnances. Apart from the short-term projects, India was also engaged in the longer-term projects in Sri Lanka such as housing, healthcare, agriculture, fisheries, education, industry, transport, culture and sports (High Commission of India in Sri Lanka n.d.). [9]

But above all, the Indian housing project in the war-affected northern Tamil population is one of the important and admirable projects in Sri Lanka. In June 2010, for the permanent rehabilitation of displaced families of northern region, Indian government announced to construct 50,000 houses in the island nation at the cost of USD 270 million making it one of the largest grant assistance projects ever undertaken by India overseas. At first phase, India entrusted an agency to construct 1,000 houses for beneficiaries in all the five districts of the Northern Province which have been complied on time. In the second phase, India initiated to construct another 45,000 houses in the Northern and Eastern Provinces. As third phase, constructions the remaining 4,000 houses in the Central and Uva Provinces for Indian Origin Tamils are in progress (Indian Housing

Project n.d.; Radhakrishnan 2012).[10] In March 2015, Prime Minister Narendra Modi visited Jaffna during his state visit to Sri Lanka and officially handed more than 27,000 new homes to displaced Tamils. India's continues support to the island nation and its public have been much appreciated by international communities (*The Indian Express* 2015; Ramakrishnan 2016).

In 13 May 2016, a tropical Cyclone Roanu which was originated in Bay of Bengal hit the Sri Lankan coast causing massive flooding and mudslides. This resulted to 201 deaths and displaced more than 1,000,000 families and damaged more than ₹ 250 billion of properties in Sri Lanka (Athas 2016; *The Indian Express* 25 May 2016). India reached out to the Sri Lankan people and assured all possible help to restore the normalcy in the island nation. Prime Minister Narendra Modi expressed his solidarity towards the affected families and directed the Indian forces to assist the Sri Lankan people during the natural disaster. India immediately dispatched two naval ships, INS Sunayna and INS Sutlej, which began sailing from Kochi with 30 tons of relief material. The relief materials include inflatable rafts, fresh water, medical supplies, clothing and other provisions necessary for disaster relief operations. Apart from providing the relief material, the Indian Navy also participated in the rescue and relief efforts. In addition, two Dornier aircrafts have been kept on standby at Kochi for any emergency requirement. An Indian Air Force C-17 Globemaster heavy transport aircraft was sent from Palam Air Force Station in Delhi with additional 60 tons of relief materials along with officials from NDMA and NDRF (Dinakar 2016; *The Indian Express* 20 May 2016).

During the recent drought in Sri Lanka, the island nation faced acquit food shortage and water crisis. It is the worst drought since 2006 and nearly 13 of the 25 districts were affected by the natural disaster. Out of 800,000 acres of agricultural land, only a little above 300,000 were planted with the staple rice crop during the last harvesting season which led to low yielding in Sri Lanka. According to the Disaster Management Centre (DMC) of Sri Lanka, approximately 709,400 people have been affected by this severe drought in Sri Lanka (*New Indian Express* 14 February 2017). Following Colombo's request, India donated 8 water bowsers and 100 metric tons of rice to Sri Lanka.[11] India's continues humanitarian assistance during the difficult times of Sri Lanka strengthened the age-old relations between India and Sri Lanka.

India's Humanitarian Assistance Programme in Maldives

Being the biggest neighbour of the island nation, India always stood by the difficult times of Maldives. During the coup d'état attempt against President Maumoon Abdul Gayoom in 1988, the Maldives government requested India's help. Following the request, Prime Minister Rajiv Gandhi immediately dispatched 1,600 troops of Fifth and Sixth Battalion of 50th Independent Parachute Regiment, to Male to dismantle the coup. Codenamed as 'Operation Cactus', Indian troops successfully disarmed the rebels and restore the Gayoom presidency at Male. Indian Navy also came to help the Maldives government, and INS Godavari and INS Betwa have been call on duty. As a result, mercenaries were chased down and arrested near the Sri Lankan coast. This prompt action by India brought international appreciation. Similarly, following the 2004 tsunami, India launched 'Operation Castor' and actively participated in the relief and rescue operation in Maldives by putting its men and machines on ground.

During the rescue operation, India launched about 44 sorties and more than 15 peoples were evacuated from outlying islands. For instance, Indian relief efforts started in Male few hours after the killer disaster and INS Mysore arrived at Male by first light on 28 December with initial relief materials (*Times of India* 27 December 2004). Similarly, two other ships INS Udaygiri and INS Aditya arrived Male with full of relief materials by 29 December. Many of these ships had integral helicopter capability and were able to provide support from the air as well, and this immediate response was invaluable in assuring the affected people that help was on the way. India's humanitarian assistance programme in Maldives during the killer tsunami was much appreciated by the island nation and enriched the bilateral relations between the South Asian countries (*Dainik Jagran* 2004; Scott 2015).

The time tested relationship between India and Maldives once again reach up to its potential during the recent water crisis in the island nation, and India went a step ahead to prove its loyalty to the friendly neighbour at the time of crisis. Following the fire incident in the only distillation plant in Maldives on 4 December 2014, India reach out to the Maldivian people and dispatched an Indian Air Force transport aircraft and naval vessels carrying a large consignment of water within hours of Maldivian request. Prime Minister Narendra

Modi and external affairs minister Sushma Swaraj called the Maldivian President Abdulla Yameen and promised full assistance to the people and leadership of Maldives (*Times of India* 5 December 2014). Subsequently two more aircrafts IL-76 and C-17 arrived in Male on the same day with 110 tonnes of drinking water to meet the need of the residents. In addition, Indian Navy's patrol vessel INS Sukanya carrying 35 tonnes of fresh water and two on board reverse osmosis (RO) plants producing 20 tonnes of fresh water per day arrived in Male and stationed to produce water round the clock to meet the water crisis. Apart from this, a large tanker vessel INS Deepak also set sail from Mumbai with 1,000 tonnes of water. In addition to the carried contingent water, it was also capable of producing 100 tonnes of drinking water every day, which might add strength to the ongoing water drinking production. These ships will stay at Male harbour until the desalination treatment plant in Male has been repaired. Two more offshore patrol vessels (OPVs) have been kept on standby, for any future requirements. India's quick and overwhelming response to Maldivian water crisis was well appreciated by the Male administration and helped to uphold the spirit of 'the close and time-tested ties', between India and Maldives.

India's Human Assistance Programme in Indonesia

In addition to the assistance to Sri Lanka and Maldives, the Indian military was able to reach out to Indonesia as well. As part of 'Operation Gambhir', the Indian Navy reach out to the Indonesian tsunami by deploying a hospital ship, providing relief supplies and setting up medical camps to aid disaster victims. INS Khukri and INS Nirupak were sent to Indonesia with 40 tonnes of relief aid including 3 tonnes of medical relief supplies. INS Nirupak which is a survey ship was converted into a full-fledged 50 bedded hospital ship equipped with medical facilities such as X-ray and ICU and dispatched to the worst affected country on the fourth day of killer tsunami, with 40-member medical team. During their stay at Meulaboh, the worst affected region of Indonesian tsunami with more than 1,800 patients were treated by the Indian forces. In addition, India was among the first countries to provide relief aid of US $1 million to Indonesia following the 2004 Tsunami.[12] India's timely help added a new chapter to age-old relationship between India and Indonesia and own international goodwills. (*Times of India* 2 January 2005).

India's Humanitarian Assistance Programme in Myanmar

Following the Cyclone Komen in August 2015, Myanmar was severely affected by widespread floods and landslides and 12 of the 14 states submerged affecting more than 1 million people and killed more than 100. In this difficult times, India reached out to its eastern neighbour and immediately sent one C-17 aircraft with 48 tonnes of rice, noodles and medicines to Mandalay (BBC 10 August 2015). In subsequent operations, India dispatched two more C-130 aircrafts and airlifted additional 55 tonnes of relief supplies from Delhi and landed in Kalay. The operation is being conducted by Ministry of External Affairs in close coordination with Government of Myanmar, Ministry of Home Affairs, Ministry of Health, Ministry of Defence, the Indian Air Force, the Embassy of India, Yangon and the Consulate of India in Mandalay. In total, Indian Air Force had air dropped about 100 tonnes of food and medicines during this hour of crisis (*Business Standard* 7 August 2015). However, the most devastated natural disaster faced by Myanmar is Cyclone Nargis in May 2008 which resulted in approximately 200,000 death fatalities and 55,000 missing people (*BBC* 7 May 2008, News). The second deadliest cyclone of all time made a landfall on the densely populated Irrawaddy Delta region with winds of 200 km per hour speed on 2 May 2008, causing the highest ever fatalities in the history of Myanmar and washing away property of USD 10 billion in the poor regions of Myanmar.[13] As per the United Nations data, about 1.50 million people were affected by the disaster cyclone (*Reuters* 8 May 2008).[14]

However, the flow of international disaster assistance was not possible as most of the Western countries and the international public opinion were critical about the ruling military junta, which shacked the democratic government in 1962. However, during this difficult time, India reached out to Myanmar and launched 'Operation Sahayata'. As a part of the operation, New Delhi sent two Indian Air Force planes which carried 4 tonnes of relief supplies each and two naval ships with more than 100 tonnes of relief material to Myanmar. In the following days, India also dispatched third consignment of 32 tonnes of relief material by an IL-76 aircraft of IAF to the neighbouring country. In another operation, two more AN-32 aircrafts, carrying relief material including tents and medicines, reached Yangon. The Indian navy also engaged in relief and rescue operation in Myanmar and two Indian Naval ships—INS Rana and INS Kirpan—carrying essential relief material such as clothing, blankets, medicine and foods reached Yangon. On 16 May 2008, India offered to send a team of 50 medical personnel to set up two independent mini-hospitals in the Irrawaddy Delta, which was accepted by the Burmese government. An aircraft carrying the team of doctors and approximately 6 tonnes of medicines was being prepared at Delhi's Palam Air Force Base.[15] New Delhi also persuaded the military junta to take international assistance, when the junta government was reluctant to allow the military ships and cargos to enter into Myanmar territories. The timely humanitarian assistance by India and its political engagement with the Myanmar government in international relief and rescue operation was applauded by international communities and by the Naypyidaw (Sharma et al. 2011).[16]

In 2015, another horrific cyclone named Komen made a landfall in Myanmar and caused severe floods and landslides in 12 regions of Myanmar. More than 100 people were killed and around 250,000 were affected by the killer disasters (*BBC* 10 August 2015). In addition, about 1.2 million acres of rice fields have been destroyed and caused widespread displacement in Myanmar. On July 31, President of Myanmar declared Chin and Rakhine States and Magway and Sagaing Regions as 'natural disaster zones' and requested for international assistance. The immediate extension of assistance to Myanmar is indicative of the strong ties of friendship between the two countries.

India was one of the states that reached out to Myanmar's request and sent 100 tonnes of medicines and food items. On 6 August 2015, two C-17 aircrafts of Indian Air Force reached Mandalay with 100 tonnes of rice, noodles and medicines. In addition, another IAF C-130 plane reached Kalay with 10 tonnes of relief supplies from Delhi on the next day.[17] In this grief moments of Myanmar, India also considered to send its men on ground and put its specialized team of NDRF on standby for assisting the relief and rescue operation in Myanmar. Addressing the press, Minister of State for Home Affairs Kiren Rijiju confirms that a team of about 200 personnel of NDRF is expected to be sent to Myanmar after India receives a formal communication (*DNA* 5 August 2015; *Zee News* 4 August 2015). India's continues support to the most disaster-prone neighbour and its disaster diplomacy brought much needed goodwill between the stained relationship between India and Myanmar.

India's Humanitarian Assistance Programmes in Afghanistan

Being one of the important neighbours and the gateway to oil rich Central and East Asian region, Afghanistan shares great cultural and economic ties with India since centuries. While Indo-Afghanistan political relations dates back to centuries, it was temporarily disrupted during the Taliban regime in 1990s. However, after the fall of Taliban, India's relations with Afghanistan were strengthen. After the long war, India played a significant role in the reconstruction and pledged USD 2 billion towards the rehabilitation process in Afghanistan. This makes India one of the leading donor nations in Afghanistan. Since 2001, India invested heavily in various medium and large projects contributed for the peace, stability and prosperity of war-turned nation.[18] Among the number of infrastructural projects, the construction of a 218 kilometre long road from Zaranj to Delaram, 220 kilovolt DC transmission line from Pul-e-Khumri to Kabul and a 220/110/20 kilovolt sub-station at Chimtala Building of Afghanistan Parliament.

During the consecutive wars with world two biggest superpowers, Afghanistan's connectivity and communicative infrastructures were severely damaged resulting in inconvenience to public life. To boost the regional connectivity, India took up the construction of 218 kilometres long road Zaranj to Delaram to the Iranian border and initiated the Chabahar Port project for facilitating movement of goods and services by passing Pakistan (*The Hindu* 25 May 2016). In addition to this, India also built 58 kilometres of inner city roads in the major cities of Afghanistan that include 40 kilometres road in Zaranj, 10 kilometres in Gurguri and 8 kilometres in connecting Gurguri to Razai.[19] To improve the city and regional connectivity of Afghanistan, India also gifted about 1,000 vehicles costing approximately US $25 million which includes 400 buses, 200 mini buses, 105 municipality utility vehicles and 285 army vehicles (*Afghan Zariza* 26 July 2015.). This recent gift of 1,000 of buses is in addition to its earlier gift of 50 buses in 2002 when Indian External Affairs Minister, Jaswant Singh, handed over the buses to Afghan officials at New Delhi (*The Hindu* 30 April 2002). To improve the telecommunication infrastructures and facilitate greater integration of the country, India also upgraded the existing telephone exchanges of 11 provinces with a total cost of about US $11.1 million and expanded the national TV network by providing an uplink from Kabul and downlinks in all 34 provincial capitals with the total cost of US $6.8 million.

During the decades of war, the power distribution system of Afghanistan was damaged drastically and Kabul, the capital city, used to get only three hours of electricity every alternate day. To improve the rusted power transmission lines, India spent US $111 million for the construction of 202 kilometre long 220 kilovolt double circuit transmission line from Pul-e-Khumri to Kabul enabling Afghanistan to import power from Uzbekistan via Pul-e-Khumri to Kabul. India also set up an electrical substation at Chimtala near Kabul and an additional substation for local electricity distribution at Charikar as its effort to boost electrification programme in Afghanistan (Chaturvedi n.d.). India also completed the Salma Dam Power Project, which is now generating of 42 megawatt electricity, besides adding an irrigation potential of 40,000 hectares of farmland in Afghanistan. To transmit the produced electricity, India also constructed a 110 kilovolt transmission line from Chishti Sharif to Herat, with 532 towers over a distance of 143 kilometres. In addition, India also supplied equipment for a 125 kilometre transmission line from Andhkhoi to Maimana, three 110/20 kilovolt sub-stations at Maimana, Faizabad and Juma Bazar, and four 20/0.4 kilovolt pole-mounted substations in Faryab Province.

To boost the agricultural productivity, India constructed the Salma Dam Power Project on river Harirud in Herat Province with a total cost of US $290 million. This dam was renamed as 'Afghan-India Friendship Dam'. This 3.7 kilometre long dam facilitates the irrigation of 75,000 hectares of farmland of 640 villages in the Herat Province (Swami 2016). India also dug 50 deep tube wells in Badgis, Baghlan, Balkh, Faryab, Herat and Jawzjan provinces between 2003 and 2007 and completed the rehabilitation of six mini hydro projects, Khanabad Irrigation Project, Qargha Reservoir and Amir Ghazi Dam to enrich Afghanistan's irrigation networks.

To strengthen the badly affected healthcare services, five Indian Medical Missions (IMMs) have been working in Kabul, Herat, Jalalabad, Kandahar and Mazar-i-Sharif, attending and disbursing medicines to 30,000 patients per month. India undertook the rehabilitation of the Indira Gandhi Institute for Child Health (IGICH) in Kabul, the largest paediatric hospital in Afghanistan, and completed its new three storied surgical block, the polyclinic block, the diagnostic block and decentralized waste water treatment system. It also provided medical equipment and artificial limbs for the war-turned victims of Afghanistan. To increase the capacity of the existing Afghan doctors, India provided them the specialists training at the All India Institute of Medical Sciences, New Delhi.

As the largest democratic state of the world, India always committed to strengthen democratic institutions in Afghanistan. India took up the construction of new parliament building in Kabul with a cost of US $83 million and handed it over to Afghan authority in 2016. India also encouraged parliamentary exchange programmes and provided the training to the Afghan Parliament Staffers in India. Indian parliamentarians also visited Afghanistan's Wolesi and Meshrano Jirgas. India also provides training to the electoral officials of Afghanistan's Election Commission.

To strengthen Afghanistan's deteriorating education system, India took several projects in Afghanistan. India contributed about US $5 million for the reconstruction of Habibia School in Kabul and committed a grant-in-aid of US $1 million to prestigious schools over the next 10 years.

India also provided 500 annual long-term university scholarships and 500 short-term diploma and technical scholarships to public servants in Indian technical and professional institutions of their choice. India helped the Afghan government for the establishment of Afghan National Agriculture Science and Technology University, Kandahar. India also helped to establish the India–Afghanistan Vocational Training Centre for training Afghan youth in carpentry, plumbing, welding, masonry and tailoring executed by the Confederation of Indian Industry. Similarly, a dedicated Women's Vocational Training Centre was established by India in Baghe-Zanana for training Afghan women. As a part of Afghanistan's 'Back to School' campaign, India supplies about 32,000 tonnes of high protein biscuits to 6.4 million Afghan school children. Every day, nearly 2 million Afghan school children in 33 out of 34 provinces (excluding Kabul) receive a packet each of 100 grams of biscuits. This provides children necessary nutrients to prevent short-term hunger and encourages school attendance. It has, thus, been a big factor in increasing school enrolment in Afghanistan.

To meet the food crisis of Afghanistan and to strengthen its strategic food reserves, India committed to supply 1.1 million metric tonnes of wheat to its northern most neighbour, of which 711,882 metric tonnes of wheat or equivalent amount (₹ 989.45 crore) in cash is already delivered. Along with these major developmental programmes, India also pledged US $1 billion to the people of Afghanistan. In addition, commitment to contribute US $1 million annually over the next five years to Afghan Red Crescent Society for the treatment of Afghan children with congenital heart disease.

A significant addition to India's development portfolio in Afghanistan is the Small Development Projects (SDP) scheme in the fields of agriculture, rural development, education, health and vocational training. The SDPs are implemented in three phases: under I and II phases (total commitment of US $20 million), 132 projects at a cost of approximately US $19.5 million were approved (94 projects completed). MoU for implementation of the third phase of the SDP scheme was signed during President Karzai's visit to India in November 2012 with an additional outlay of US $100 million. A total of 287 projects at approximately US $52.5 million have been approved under Phase III and are at various stages of implementation. Eighty-seven additional projects are under GoI's active consideration. India's unprecedented humanitarian assistant programme earned much appreciation from the people of Afghanistan and earned huge respect from international community.

India's Humanitarian Assistance Programmes in Philippines

India and Philippines share a friendly and warm relation since centuries. The cultural and economic relations between the two countries date back to second century AD, when Buddhism and bilateral trades were flourishing. As a part of the British military expedition, about 600 British Indian Army reached Philippines to take part in various Anglo-Spanish wars that continued from 1762 to 1764. After the withdrawal of the British troops, many of the Indian troops chose to settle in Philippines, creating Indian diaspora in Philippines (Santarita 2015). However, the post-independence relations between the two countries started deteriorating due to the Cold War politics, where both the countries perceived themselves juxtapose in the rivalry power blocks. Though the Cold War got over long time back, India's relation with Philippines is limited to occasional official visits and harbouring of Indian naval ships.

However, when a bad hurricane, Haiyan, affected Philippines in 2013, India reached out to the people of Philippines and extended all possible assistance to people in need. The deadly typhoon, with wind speed of 380 kilometre per hour, crossed over most part of Philippines causing unprecedented damages to the life and properties in 60 of the 80 provinces. As per the government data, more than 6,000 people have been killed and approximately 1,800 were reportedly missing. The subsequent heavy pouring caused floods and landfalls in most of the provinces adding further grief to the affected people (*CNN* 13 December 2013).

In this difficult times of humanity, India reached out to the Philippine government and sent the C-130 transport aircraft with 15 tonnes of relief material which included medicines, fresh water, water purification materials, mattresses, food packets and milk powder. Prime Minister Manmohan Singh while dispatching the relief materials stated that India stands by the people of Philippines and conveyed that 'We stand ready to provide all help to support the relief and rehabilitation efforts of your government'. India also send a naval ship with additional relief materials, including water treatment machinery, house-building materials and other necessary items for relief and rehabilitation of affected people (*Times of India* 13 November 2013). In past, India had also reached out to Philippines when typhoon Durian made the landfilled in Philippines in 2006 that killed more than 1,300 people and left hundreds more missing. During this natural calamity, India provided necessary relief material to Philippines. India's continues humanitarian assistance to the disaster-prone nation earn much international appreciation and strengthened its soft power internationally.

India's Humanitarian Assistance Programmes in Pakistan

The relationship between India and Pakistan has been marked hostile and difficult since the partition of the subcontinent in 1947. Time and again, both the states have been engaged in several small- and large-scale wars making South Asia most disintegrated region of the world. Due to the prolonged hostility, there was less scope for any kind of cooperation between the two neighbours including at the time of natural disaster. Being situated in South Asia, one of the most disaster-prone regions of the world, both the countries faced numbers of deadly natural disasters repeatedly. While Pakistan witnessed major natural disasters in recent times such as the 2005 Kashmir earthquake, 2007 Cyclone Yemyin, 2010 Pakistan floods and

2015 Kashmir floods; India also suffered affected periodical natural calamities including 1999 Odisha cyclone, 2001 Gujarat earthquake, 2002 Indian heat wave, 2004 Indian Ocean tsunami, 2007 Bihar flood, 2005 Mumbai catastrophes, 2010 Eastern Indian storm, 2013 Maharashtra drought, 2013 Uttarakhand flash floods and 2014 Kashmir floods. Despite witnessing common disasters across the border and suffering equally, there is hardly any visible disaster cooperation between India and Pakistan.

However, in recent time, both the neighbours showed their sympathy on each other's sufferings and extended some degree of cooperation in the difficult times of humanity in the subcontinent. For instance, during the 2005 Kashmir earthquake India and Pakistan have cooperated each other in the relief and rescue operations. During this devastated disaster people across the border suffered and approximately 55,300 people died due to the killer earthquake. The earthquake was of 7.6 in magnitude and epicentre was near Muzaffarabad, the capital of the Pakistan administered Kashmir. The Pakistan side was severely damaged and more than 54,000 people died compared to the number of people died on the Indian side. In this difficult time which was more than 1,300. India expressed its sympathy with the affected people of Pakistan administered Kashmir. Indian Prime Minister Manmohan Singh offered assistance to Pakistan and immediately dispatched 25 tonnes of relief material to Pakistan including food, blankets and medicine. On October 12, additional seven truckloads (about 82 tons of stores) of army medicines, 15,000 blankets and 50 tents were delivered to Pakistan authority through an Ilyushin-76 cargo plane that flew across the border (*The Dawn* 13 October 2005). Two days later, India sent another consignment of relief materials to Pakistan, by train through the Wagah Border. The consignment includes 5,000 blankets, 370 tents, 5 tons of plastic sheets and 12 tons of medicine. India also dispatched another consignment of medicine and relief material by train. Along with the necessary relief materials, India also pledged $25 million as foreign aid to Pakistan for the rehabilitation of quake victims (*BBC* 27 October 2005). In addition, Indian software giant Infosys also offered an aid of US $226,000 to the Pakistan government (Banerjee n.d.). To facilitate the relief and rescue operation, India opened the three border points on the Line of Control (LoC). This was the first time in decades that India has offered any disaster aid for Pakistan.

Similarly, during the devastated floods in Pakistan in July 2010, which made severe damage to the life and property of the four populated regions of the Islamic republic, killing approximately 2,000 people and affecting another 20 million, India reached out to its Western neighbour conveying its sympathy to the affected people. Prime Minister Manmohan Singh called to his Pakistani counterpart Yusuf Raza Gilani and ensured all possible help to Pakistan. India also offered US $5 million as relief aid initially and increased to US $ 5 million later and become the top 10 donor country on Pakistan flood. China, Pakistan's all time friend, however responded half-heartedly and committed only US $9 million, which has been largely criticized by Pakistani media (Joshua 2010; *Times of India* 21 August 2010).

In 2014, the late monsoon caused severe flood in both sides of the Kashmir resulting in the death of 277 people in India and about 280 in Pakistan (*Zee News* 23 September 2014). While managing its own relief work, India offered all possible help to Pakistan. Prime Minister Narendra Modi wrote to his Pakistan counterpart Nawaz Sharif offering Indian assistance in the relief and rescue mission in Pakistan. While showing sympathy towards the flood victims, he wrote that 'my heart goes out to the affected people and my deepest sympathies are with them and their families. In this hour of need, I offer any assistance that you may need in the relief efforts that will be undertaken by the Government of Pakistan. Our resources are at your disposal wherever you need them' (*The Indian Express* 7 September 2014). India kept several medical teams and other relief workers on standby and nearly 400 Indian medical staff have been waiting for the Pakistan government's visa approval to help flood victims (Agarwal 2014). Though Pakistan did not ask any assistance from India in the Kashmir flood, India's *altruistic* gesture sends positive image towards the world communities. Not only in natural disaster, India also proposed to help Pakistan in eradicating polio, as it has done in their side of the border (*The Hindu* 25 October 2014). As per the experts, these gestures will ease the tensions and boost the level of cooperation between the two neighbours.[20] Though the chance of lasting cooperation between India and Pakistan apparently is slim, India humanitarian assistance efforts towards its historic rivalry shows that when it comes to humanitarian aid, it does not think twice before reaching out to the neighbours.

India's Humanitarian Assistance Programme in Bangladesh

Once a part of Indian state Bangladesh shares historic ties with India. The whole of Bangladesh, as a part of undivided Bengal, remained an integral part of India till 1905 and played an important role in shaping India's sociocultural identity. However, as a part of Mountbatten Plan, the Muslim majority East Bengal was awarded to Pakistan and renamed as East Pakistan after independence. Following the language movement in 1954 and the subsequent Bengali uprising against Pakistan government, India intervened and sent its army to East Pakistan and facilitated to establish an independent state of Bangladesh in 1971. Since then India gives unprecedented support to Bangladesh and stands by every difficult moments of its eastern neighbour. Even during the 1971 war, India offered food and shelter to approximately 10 million Bengali refugees, those fled to India from East Pakistan (now Bangladesh). To address the problems of the refugees, India appointed a cabinet minister along with the Minister for Labour and Rehabilitation. India spent around ₹ 100 million on food, clothing and shelter for the refugees in the first few months after the war. India also provided emergency medical facilities for the war-turned refugees and provided necessary vaccines and antibiotics during the outbreak of cholera disease in the refugee camps. At the end of the war, India also allowed approximately 1.5 million refugees to stay back after Bangladesh became independent (Price 2005). In another instance approximately 14,214 Bangladeshi citizens were given Indian citizenship with the implementation Indo-Bangladesh Land Boundary agreement on 31 July 2015 (Sen and Bhattacharya 2015).

Apart from providing shelter to refuges, India also carried out a number of humanitarian assistance programmes in the difficult times of Bangladesh. And, there are no bigger difficult times for the young nation than the natural disasters. Being the sixth most risk-prone country of the world, Bangladesh faces frequent natural calamities which killed millions of people since independence. Since 1907, this region witnessed about 212 major natural calamities, including 137 cyclones, 64 floods, 6 earthquakes and 5 droughts. In these major disasters, approximately 5,776,221 people have lost their lives.[21] Despite its difficult relationship with Bangladesh after independence, India stood with its eastern neighbour during the natural disasters time and again. For instance, during 1991 cyclone in Bangladesh that killed 138,866 people and destroyed the property worth USD $1.7 billion, India sent the relief materials such as medicine, foods, tents and blankets worth USD 1 million to Bangladesh. Prime Minister Chandra Shekhar visited Bangladesh on 12 May 1991 and assured all possible help to affected people. India provided 10 tons of rice to Bangladesh and put 6 Indian Air Force helicopters for relief and rescue operation in the cyclone-affected areas. India also sent a team of railway officials to excess the damage caused by the typhoon and assist the Bangladesh government for the restoration of rail services. However, comparatively, India's disaster aid was very small due to its difficult relationship with its eastern neighbour and severe domestic financial crisis has lot to contribute towards this meagre amount of aid to Bangladesh (Singh 2010).[22]

However, the estranged relationship started to get back to normalcy ever since Sheikh Hasina came to power. Since then, both the South Asian nations have been cooperating host of issues including disaster management. For instance, following the cyclone Sidr, India sent necessary relief materials including 40,000 megatonnes of rice, 1,000 megatonnes of skimmed milk powder, medicines, tents, blankets, ready-to-eat meals and portable water purifiers to Bangladesh. Four ships of Indian Navy, INS Gharial, INS Mahish, INS Cheetah and INS Kumbhir, put in service to transport the relief materials to Chittagong Port of Bangladesh. In addition, New Delhi also committed financial aid worth more than 250 crore taka (more than US $37 million) to carry out the reconstruction programme in the cyclone-affected areas. Indian External Affairs Minister Shri Pranab Mukherjee visited the cyclone-hit areas in December 2007 and expressed India's sympathy to the affected people of Bangladesh in their hour of need. India also engaged directly for the rehabilitation of ten worst cyclone-affected villages and constructed more than 2,800 core shelters in the affected villages in Bagerhat district in southern Bangladesh. Similarly, 2,800 solar lamps have also been handed over to the beneficiaries of these core shelter houses.[23]

Located in the same geographic region and witnessing similar natural calamities, India and Bangladesh also decided to coordinate and cooperate in disaster management. As a part of 'South Asian Annual Disaster Management Exercise' (SAADMEx), disaster relief agencies of India and Bangladesh have been conducting several joint drills and learned each other's skills to reduce the disaster fatalities.[24] Both the armed forces also conducted joint military exercises in disaster relief and rescue operations (Habib 2016). During the occasion of 'Asian Conference on Disaster Risk Reduction', a Bangladeshi delegation visited New Delhi under the leadership of Disaster Management and Relief Minister Mofazzal Hossain Chowdhury Maya and signed the bilateral agreement on disaster management with Government of India. By signing the agreement, both the countries agreed to cooperate with each other in disaster prepended and relief operations.

Besides disaster cooperation, India is also engaged in various humanitarian activities in Bangladesh. To assist the developmental projects, India offered the first Line of Credit of US $1 billion to Bangladesh in 2010. India also announced additional aid of USD 2 billion to its eastern neighbour during the visit of the Prime Minister Narendra Modi to Bangladesh in June 2015, making the single largest credit line offered by India to any country so far (Loiwal 2016). Apart from financial aid, India is involved in number of projects in Bangladesh to improve the life style of the people. To improve the power shortage of its neighbour, India provided 500 megawatts electricity by establishing inter-grid connectivity in 2013. In 2016, an additional 100 megawatts of power supplied to Bangladesh by inter-connecting Comilla-Tripura link. In addition to the 600 megawatt electricity supply, India also assist the Bangladesh government in setting up a 1320 megawatt coal based power project at Rampal, Bangladesh. India's National Thermal Power Corporation (NTPC) signed an MoU with Bangladesh Power Development Board (BPDB) on 30 August 2010 and entered into a 50:50 partnership joint venture to complete the project. Similarly, India's two private farms proposed to set up large power plants in Bangladesh with a total capital investment of US $5.5 billion. While Reliance Power agreed to set up a 3,000 megawatt LNG-based power plant with a total cost of US $3 billion, the Adani Power proposed to construct a 1600 megawatt coal-fired power plant at a cost of US $1.5 billion in Bangladesh during Prime Minister Modi's visit to Bangladesh (*DNA* 6 June 2015). To strengthened Bangladesh's renewable energy capacity, India signed the MoU on 'India-Bangladesh Renewable Energy Cooperation' on 6 September 2011 which enable technical and institutional cooperation between the two countries. To improve the healthcare facility of Bangladesh, both the countries also signed agreement to cooperate in healthcare with exchange of doctors and health professionals and carry out joint research in medical science. To encourage the education and research activities in Bangladesh, India provides ICCR scholarships to 200 Bangladeshi students every year. India has offered scholarships for meritorious Bangladeshi students and PhD researchers to undertake studies in various professional courses including medical science. To improve the ageing railway networks of Bangladesh, India provided railway equipment including locomotives and passenger coaches to Bangladesh government (Srivastava Balaji 2016). New Delhi also signed the agreement to construct the Akhaura–Agartala rail link with a cost of INR 972.52 crore to improve connectivity and boost trade between the two countries (*The Indian Express* 20 October 2016).

Conclusion

Being placed at the most disaster-prone region of the world, South Asia faces periodical disasters and suffered heavily due to the natures furry. The rapid growth of the South Asian population and its increased concentration in few places often in hazardous environments has escalated both the frequency and severity of natural disasters. With the tropical climate and unstable land forms, coupled with deforestation, unplanned growth proliferation, non-engineered constructions which make the disaster-prone areas more vulnerable, tardy communication, poor or no budgetary allocation for disaster prevention, South Asian countries suffer more or less chronically by natural disasters. Starting from flood to drought, tsunami to earthquake, this region witnessed all types of natural disasters that includes 1999 Odisha super cyclone, 2001 Gujarat earthquake, 2004 Indian Ocean tsunami, 2005 Kashmir earthquake, 2007 Bangladesh cyclone, 2008 Baluchistan earthquake, 2010 Pakistan floods, 2013 Uttarakhand flood, 2014 Kashmir floods and 2015 Nepal earthquake.

Most of the natural disasters have cross border impact and affected multiple nations at the same time. Traditionally, South Asian states are largely dependent on unilateral approach in relief and rescue operation and largely failed to give quick relief to the disaster victims. Lack of confidence and prolonged trust deficit between the states of this region led to the absence of any multilateral and regional disaster response mechanism. In addition, due to slow economic growth and lack of financial muscle, the neighbouring countries have very little to share with the victim country at the time of disaster. However, the situation started changing after the economic transformation of India and its rise as a regional power. Soon, India started participating in several disaster rescue and relief operations in its neighbourhood and funded many developmental activities of the regions. Though India has been active in providing financial and technological cooperation towards its neighbours since independence, the nature and volume has drastically increased in recent years. From 2004 Indian Ocean tsunami to recent Nepal earthquake, time and again India stood with its neighbours at the time of natural disasters and extended all possible assistance to the victim countries.

Though India's disaster efforts during recent Nepal earthquake got the world attention, New Delhi has been engaged in trans-national relief operations in the past. India's offer of help Pakistan during earthquake in 2005; its assistance during 2004 tsunami in Sri Lanka, Bangladesh and Maldives; and various relief efforts in monsoon flood and deadly cyclones in Bangladesh, Sri Lanka and Myanmar have earned much goodwill for India and enhanced its soft power image in the neighbourhood. However, the magnitude of India's relief programme was low due to its economic and technical resources. However, it has today shown the world that it can handle crisis like the one she faced in Yemen in evacuating its beleaguered citizen from an active war zone and in Nepal where massive humanitarian assistance was carried out by India. Hence, it is essential to discuss, in detail, India's disaster assistance programme in the neighbourhood.

There is no denying that the Indian assistance to these nations is modest given the scale of the relief that is required but the spontaneity with which it was extended and the alacrity with which the Indian military was able to arrive are indicative of the credibility of the Indian humanitarian response machinery that spans the political, diplomatic and military determinant and this has some pertinent pointers for the regional security grid that the professionals will be monitoring.

India is still in the process of defining its role in disaster relief efforts. Since its independence, the country has come to the aid of people in need. For example, it became home for thousands of Tibetan Refugees in 1959 and millions from East Pakistan (now Bangladesh) in 1971. In 1987, during the Sri Lankan Civil War, India flew humanitarian assistance materials to civilians in the city of

Jaffna, an act that could be viewed as one of the few humanitarian interventions worldwide. In the past 10 years, the scale and frequency of India's efforts to help those in distress have changed significantly. Today, as the world's fourth largest economy, India has the means to contribute to international aid efforts more systematically. Although a large share of its population is still poor and huge income gaps characterize the Indian economy, India has come a long way; once dependent on Western aid, it is now part of the group of non-Western humanitarian donors which together account for at least 12 per cent of worldwide humanitarian aid each year (Harmer and Cotterrell 2005: 16). The Indian government has time and again proved that when it comes to humanitarian aid, it does not think twice before reaching out to the neighbours.

Notes

1. "Why Is South Asia Vulnerable to Climate Change?" World Bank. 2009. Available at http://web.worldbank.org/WBSITE/EXTERNAL/COUNTRIES/SOUTHASIAEXT/0,,contentMDK:22404173~pagePK:2865106~piPK:2865128~theSitePK:223547,00.html
2. "South Asia: Shared View on Development and Climate Change, South Asia Region Sustainable Development Department". Washington, DC: Word Bank, 2009. p. 4.
3. Data comprised from "World Risk Report 2016". Institute for Environment and Human Security, United Nations University, Bonn. Available at https://collections.unu.edu/eserv/UNU:5763/WorldRiskReport2016_small.pdf
4. Ibid.
5. Ibid.
6. "Developmental Projects in Sri Lanka". High Commission of India in Sri Lanka. Available at http://www.hcicolombo.org/partner.php?id=209
7. "Demining in Northern and Eastern Provinces" (n.d.). High Commission of India in Sri Lanka. Available at http://www.hcicolombo.org/pdf/1_demining.pdf
8. Developmental Projects in Sri Lanka.
9. Ibid.
10. Indian Housing Project (n.d.). High commission of Indian, Sri Lanka, Available at http://www.hcicolombo.org/pdf/TENDER_33.pdf

11. "India to Donate 8 Water Bowsers to Drought-hit Sri Lanka". February 21, 2017. Available at http://www.business-standard.com/article/news-ians/india-to-donate-8-water-bowsers-to-drought-hit-sri-lanka-117022101087_1.html
12. Press release, Embassy of India, Jakarta. Available at http://indianembassyjakarta.com/index.php?option=com_content&view=article&id=434
13. "Asian Bloc to Handle Burma Aid". the star.com, May 19, 2008. Available at https://www.thestar.com/news/world/2008/05/19/asian_bloc_to_handle_burma_aid.html
14. "Myanmar: Cyclone Nargis 2008 Facts and Figures, The International Federation of Red Cross and Red Crescent Societies (IFRC)", May 3, 2011. http://www.ifrc.org/en/news-and-media/news-stories/asia-pacific/myanmar/myanmar-cyclone-nargis-2008-facts-and-figures/
15. "Indian Relief Material Reaches Myanmar". May 8, 2008. Available at http://www.rediff.com/news/report/myan/20080508.htm
16. "Myanmar's Cyclone Crisis and India's Aid". *The Hindu*, May 30, 2008. Available at http://www.thehindu.com/todays-paper/tp-opinion/Myanmarrsquos-cyclone-crisis-and-Indiarsquos-aid/article15231883.ece
17. "Indian Air Force Sends Relief Material to Myanmar". Press Information Bureau, Government of India, August 7, 2015. Available at http://pib.nic.in/newsite/PrintRelease.aspx?relid=124634; "Disaster Relief Support to Myanmar in Response to Flood Emergency". Press Release, Ministry of External Affairs, August 6, 2015. Available at http://mea.gov.in/press-releases.htm?dtl/25655/Disaster+Relief+Support+to+Myanmar+in+response+to+Flood+Emergency
18. India–Afghanistan Relations, Embassy of India in Afghanistan. Available at http://eoi.gov.in/kabul/?0354?000
19. "India-Afghanistan: A Partnership". Report published by Embassy of India in Kabul, Public Diplomacy Division, Ministry of External Relations, Government of India. Available at http://eoi.gov.in/kabul/?pdf0358?000
20. See Dipankar Banerjee and G. Parthasarathy's opinion about India–China disaster cooperation. Vibhuti Agarwal, "India, Pakistan Engage in Flood-Aid Diplomacy". *The Wall Street Journal*, September 8, 2014. Available at http://blogs.wsj.com/indiarealtime/2014/09/08/india-pakistan-engage-in-flood-aid-diplomacy/
21. "Bangladesh Natural Disaster Profile". Centre for Hazards and Risk Research, University of Colombia. Available at https://www.ldeo.columbia.edu/chrr/research/profiles/pdfs/bangladesh_profile1.pdf

22. See also, "India-Bangladesh Relations". Ministry of External Affairs, Government of India. Available at https://www.mea.gov.in/Portal/ForeignRelation/Bangladesh_July_2014_.pdf
23. Ibid.
24. "First Time: India to Host SAARC Disaster Management Exercises". *The Economic Times*, November 5, 2015. Available at http://economictimes.indiatimes.com/news/politics-and-nation/first-time-india-to-host-saarc-disaster-management-exercises/articleshow/49675667.cms.

References

Afghan Zariza, "India to Provide 1,000 New Millie Buses to Afghanistan and Develop Three Workshops in Kabul". July 26, 2015. Available at http://www.afghanzariza.com/2015/07/26/india-to-provide-1000-new-millie-buses-to-afghanistan-and-develop-three-workshops-in-kabul#sthash.PRCNzN8B.dpuf

Agarwal, Vibhuti (2014). "India, Pakistan Engage in Flood-Aid Diplomacy". *The Wall Street Journal*, September 8, 2014. Available at http://blogs.wsj.com/indiarealtime/2014/09/08/india-pakistan-engage-in-flood-aid-diplomacy/

An Lu (2005). "China Offers 20 mln Yuan in Humanitarian Aid to Quake-Hit Nepal". *Xinhua News*, April 26, 2015. Available at http://news.xinhuanet.com/english/2015-04/26/c_134186099.htm

Athas, Iqbal (2016). "Sri Lanka: Over 100 Dead Following Torrential Rain". *CNN*, May 25, 2016. Available at http://edition.cnn.com/2016/05/22/asia/sri-lanka-flooding-deaths/

Banerjee, Dipankar (n.d.). "Disaster Diplomacy in Asia—An Indian Perspective". Available at http://www.disasterdiplomacy.org/pb/banerjee2008.pdf

BBC, "India Offers Pakistan $25m in Aid". October 27, 2005. Available at http://news.bbc.co.uk/2/hi/south_asia/4381982.stm

BBC, "Myanmar Flooding Affects One Million". August 10, 2015. Available at http://www.bbc.com/news/world-asia-33844076

BBC, "US Presidents in Tsunami Aid Plea". January 3, 2005,. Available at http://news.bbc.co.uk/2/hi/asia-pacific/4143459.stm#map

Bhaskar, Uday C. (8 January 2005). "Tsunami Reveals Indian Military's Humanitarian Response Capability". IDSA Comment. Also Published in *Dainik Jagran*, New Delhi, on January 6, 2004. Available at http://www.idsa.in/idsastrategiccomments/TsunamiRevealsIndianMilitarysHumanitarianResponseCapability_CUBhaskar_080105

Bhatiya, Neil (2014). "Why South Asia Is So Vulnerable to Climate Change, Foreign Policy". April 22, 2014. Available at http://foreignpolicy.com/2014/04/22/why-south-asia-is-so-vulnerable-to-climate-change/

Business Standard, "India Sends in Relief to Flood Ravaged Myanmar". August 7, 2015. Available at http://www.business-standard.com/article/pti-stories/india-sends-in-relief-to-flood-ravaged-myanmar-115080701454_1.html

Chattopadhyaya, Haraprasad (1994). *Ethnic Unrest in Modern Sri Lanka: An Account of Tamil-Sinhalese Race Relations*, MD Publications. p. 102.

Chaturvedi, S. K. (n.d.). "Lighting Up Kabul". *Times of India*. Available at http://epaper.timesofindia.com/Default/Layout/Includes/ETNEW/ArtWin.asp?From=Archive&Skin=ETNEW&BaseHref=ETM%2F2010%2F03%2F29&ViewMode=HTML&EntityId=Ar01900&AppName=1

CNN, "Typhoon Haiyan Death Toll Tops 6,000 in the Philippines". December 13, 2013. Available at http://edition.cnn.com/2013/12/13/world/asia/philippines-typhoon-haiyan/

Data compiled from the Centre for Research on the Epidemiology of Disasters (CRED) Database, http://www.emdat.be/advanced_search/index.html (Accessed on March 29, 2017).

DNA, "Anil Ambani to Invest US $3 Billion in Power Plant in Bangladesh". June 6, 2015. Available at http://www.dnaindia.com/money/report-anil-ambani-to-invest-us-3-billion-in-power-plant-in-bangladesh-2092887

DNA, "India Plans to Send NDRF Teams to Myanmar". New Delhi, August 5, 2015. Available at http://www.dnaindia.com/india/report-india-plans-to-send-ndrf-teams-to-myanmar-2111218

DNA, "Operation Maitri: India launches massive relief and rescue efforts in Nepal", New Delhi 26 Apr 2015. http://www.dnaindia.com/world/report-operation-maitri-india-launches-massive-relief-and-rescue-efforts-in-nepal-2080936

Gilbert, Marc Jason (2017). *South Asia in World History*. New York; Oxford University Press. p. IX.

Habib, Haroon (2016). "India, Bangladesh Joint Exercise from Tomorrow". November 4, 2016. Available at http://www.thehindu.com/todays-paper/tp-national/India-Bangladesh-joint-exercise-from-tomorrow/article16088345.ece

Haidar, Suhasini (2016). "Chabahar and the Afghan Roundabout". *The Hindu*, May 25, 2016. Available at http://www.thehindu.com/opinion/op-ed/chabahar-and-the-afghan-roundabout/article8641602.ece

Harmer, Adele and Cotterrell, Lin (2005). "Diversity in Donorship the changing Landscape of Official Humanitarian aid Researched", Research paper, the Humanitarian Policy Group, September 2005. Available at https://www.odi.org/sites/odi.org.uk/files/odi-assets/publications-opinion-files/275.pdf.

Joshua, Anita (2010). "Pakistan Flood: India among Top 10 Aid-givers". *The Hindu*, August 31, 2010. Available at http://www.thehindu.com/news/international/Pakistan-flood-India-among-top-10-aid-givers/article15898053.ece

Lancaster, John (2005). "India Takes Major Role in Sri Lanka Relief Effort". *Washington Post*, January 20, 2005. Available at http://www.washingtonpost.com/wp-dyn/articles/A22194-2005Jan19.html?nav=rss_world/asia/southasia/srilanka

Loiwal, Manogya (2016). "Bangladesh, India Sign $2 Billion New Line of Credit". *India Today*, March 9, 2016. Available at http://indiatoday.intoday.in/story/bangla-desh-india-sign-2-billion-new-line-of-credit/1/616214.html

New Indian Express, "Over 700,000 People Affected by Drought in Sri Lanka". February 14, 2017. Available at http://www.newindianexpress.com/world/2017/feb/14/over-700000-people-affected-by-drought-in-sri-lanka-1570544.html

Peri, Dinakar (2016). "India Steps in to Assist Sri Lanka in Cyclone Relief". *The Hindu*, New Delhi, May 20, 2016. Available at http://www.thehindu.com/news/national/India-steps-in-to-assist-Sri-Lanka-in-cyclone-relief/article14330007.ece

Petroff, Alanna and Rooney, Ben (2015). "Nepal Earthquake Donations: Who's Sending What". *CNN*, April 28, 2015. Available at http://money.cnn.com/2015/04/27/news/nepal-earthquake-donations/

Price, Gareth (2005). "India's Official Aid Programme", in *Diversity in Donorship: The Changing Landscape of Official Humanitarian Aid*. Background Paper, Humanitarian Policy Group, Overseas Development Institute, September 2005. Available at https://www.odi.org/sites/odi.org.uk/files/odi-assets/publications-opinion-files/416.pdf

Radhakrishnan, R. K. (August 15, 2012) "Indian Housing Project for Tamils in Sri Lanka on Track". Updated on September 6, 2016. Available at http://www.thehindu.com/news/international/indian-housing-project-for-tamils-in-sri-lanka-on-track/article3774712.ece

Ramakrishnan, T. (2016). "A House for Every Hill-Country Tamil". *The Hindi*, May 17, 2016. Available at http://www.thehindu.com/opinion/columns/A-house-for-every-hill-country-Tamil/article14322964.ece

Reuters, "UN Says 1.5 Million People Affected by Myanmar Storm". May 8, 2008. http://uk.reuters.com/article/idUKSP48798._CH_.242020080508

Richardson, John (2005). *Paradise Poisoned: Learning about Conflict, Terrorism, and Development from Sri Lanka's Civil Wars*. International Center for Ethnic Studies. p. 102.

Santarita, Joefe (2015). "Migration and Business: A Survey of Indian Communities in the Philippines", *Asian Studies: Journal of Critical Perspectives on Asia*, Vol. 51, No. 1, pp. 75–97.

Scott, David (2015). "Indian Ocean as India's Ocean", in *Oxford Handbook on Indian Foreign Policy*, ed. David M. Malone, C. Raja Mohan, and Srinath Raghavan. Oxford: Oxford University Press. p. 477.

Sen, Saibal and Bhattacharya, Pinak Priya (2015). "History at Midnight: India-Bangladesh Swap Enclaves, Over 14000 Bangladeshis Become Indians". *India Times*, August 1, 2015. Available at http://www.indiatimes.com/news/world/history-at-midnight-indiabangladesh-swap-enclaves-over-14000-bangladeshis-become-indians-243691.html

Sharma, Reetika, Goria, Ramvir, and Mishra, Vivek (2011). *India and the Dynamics of World Politics: A book on Indian Foreign Policy*. New Delhi: Pearson Education India. p. 215.

Singh, Danvir (2015). "India's Prowess and Disaster Diplomacy". *Indian Defence Review*, April 29. Available at http://www.indiandefencereview.com/news/indias-prowess-and-disaster-diplomacy/

Singh, Raj Kumar (2010). *Relations of NDA and UPA with Neighbours*. Gyan Publishing House. pp. 254–255.

Srivastava Balaji, Madhumita (2016). "India- Bangladesh Relations: An Enduring Economic Partnership". Occasional Paper, January 2016, Vivekananda International Foundation, New Delhi. Available at http://www.vifindia.org/sites/default/files/india-bangladesh-relations-an-enduring-economic-partnership.pdf

Swami, Praveen (2016). "At Afghan Dam Inauguration, PM Promises: India Will Not Forget You". *The Indian Express*, June 5, 2016. Available at http://indianexpress.com/article/india/india-news-india/narendra-modi-afghanistan-salma-dam-inauguration-ashraf-ghani-2834106/

The Dawn, "Indian Troops Cross LoC to Back Up Relief Efforts". October 13, 2005.

The Guardian, "The Human and Financial Cost of the Indian Ocean Tsunami—Interactive". December 25, 2014. Available at https://www.theguardian.com/global-development/ng-interactive/2014/dec/25/human-financial-cost-indian-ocean-tsunami-interactive

The Hindu, "India Offers to Help Pakistan Eradicate Polio". October 25, 2014. Available at http://www.thehindu.com/news/national/india-offers-to-help-pakistan-eradicate-polio/article6531827.ece

The Hindu, "Indian Buses for Afghanistan". April 30, 2002. Available at http://www.thehindu.com/2002/04/30/stories/2002043001641200.htm

The Indian Express, "Agartala-Akhaura Rail Link to Be Completed in 2.5 years". October 20, 2016. Available at http://indianexpress.com/article/india/india-news-india/agartala-akhaura-rail-link-to-be-completed-in-2-5-years-3091360/

The Indian Express, "India Rushes Two Naval Ships and C-17 Aircraft to Cyclone-hit Sri Lanka". May 20, 2016. Available at http://indianexpress.com/article/india/india-news-india/sri-lanka-floods-india-rushes-two-naval-ships-and-c-17-aircraft-to-cyclone-hit-lanka-2810941/

The Indian Express, "PM Modi Visits Jaffna, Presents 27,000 New Homes to Tamils Displaced in Civil War". March 14, 2015. Available at http://indianexpress.com/article/world/neighbours/modi-visits-jaffna-presents-27000-new-homes-to-tamils-displaced-in-civil-war/

The Indian Express, "PM Modi Writes to Nawaz Sharif, Offers Help for Flood Victims". September 7, 2014. Available at http://indianexpress.com/article/india/india-others/pm-modi-writes-to-nawaz-sharif-offers-help-for-flood-victims/

The Indian Express, "Sri Lanka Floods: Death Toll Reaches 101 as Waters Recede". May 25, 2016. Available at http://indianexpress.com/article/world/world-news/sri-lanka-floods-death-toll-reaches-101-as-waters-recede-2817926/

Times of India, "India to Send Relief to Super Typhoon-hit Philippines". November 13, 2013. Available at http://timesofindia.indiatimes.com/world/rest-of-world/India-to-send-relief-to-Super-Typhoon-hit-Philippines/articleshow/25670632.cms

Times of India, "Maldives Hit by Water Crisis, India Sends Help". December 5, 2014. Available at http://timesofindia.indiatimes.com/india/ Maldives-hit-by-water-crisis-India-sends-help/articleshow/45385033.cms

Times of India, "Navy Launches Operation Gambhir in Indonesia". January 2, 2005. Available at http://timesofindia.indiatimes.com/india/Navy-launches-Operation-Gambhir-in-Indonesia/articleshow/978531.cms

Times of India, "Pak Accepts Indian Aid after US Prod". August 21, 2010. Available at http://timesofindia.indiatimes.com/world/pakistan/Pak-accepts-Indian-aid-after-US-prod/articleshow/6369785.cms

Times of India, "Relief Ops 'Castor' and 'Rainbow' Launched". December 27, 2004. Available at http://timesofindia.indiatimes.com/india/Relief-ops-Castor-and-Rainbow-launched/articleshow/972740.cms

Zee News, "India Plans to Send NDRF Teams to Myanmar". August 4, 2015. Available at http://zeenews.india.com/news/india/india-plans-to-send-ndrf-teams-to-myanmar_1641751.html

Zee News, "J&K Floods: Death Toll Mounts to 284". September 23, 2014. Available at http://zeenews.india.com/news/jammu-and-kashmir/jk-floods-death-toll-mounts-to-284_1474762.html

PART V

Conclusion

14

Natural Disaster and Women

Towards a Sustainable Disaster Risk Reduction

Pradeep Kumar Parida

> It is not enough to understand how science has been made; we need to develop more practical, constructive ways to employ tools of gender analysis in creating what I will call 'sustainable science'. Only when gender analysis becomes an integral part of science research programs will the problem of women in science be solved.
>
> (Londa Schiebinger, 1997: 203)

> Human vulnerability—those circumstances that place people at risk while reducing their means of response or denying them available protection—becomes an integral concern in the development and evaluation of disaster policies. We must change the policies of today that rely heavily on sending assistance only after tragedy has occurred.
>
> (L. Comfort et al., 1999: 39)

The natural hazards that trigger disasters do not select its victims. To consider the complex consequences of disaster risks, the social production of vulnerability needs to be understood, more particularly, with reference to how society creates the conditions in which people face hazards differently. So far, most of the disaster victims both in developed and developing or Third World countries are already socially, economically and politically disadvantaged. In many cases, untenable political economy keeps some disadvantaged groups more at risk than others. If this were perpetually to be the case for each probable disaster, it would transpire a group or area of society that is already disadvantaged compared to other groups and areas. Therefore, disaster preparation and response should be 'need-based'. Moreover, within particular socially disadvantaged groups, everyone does not necessarily suffer equal disadvantages in disasters. Thus, need and requirement for medicines, foods, water, house, transport for evacuation or personal mobility assistance must translate directly into preparation and resource deployment in egalitarian ways. Nevertheless, my intention here is to put forth a larger issue with regard to sustainable disaster reduction.

Given that the setting of this unit is sustainable DRR, examining disaster through a gender lens would reveal significant differences in the ways women and men respond to hazards as well as to the survival challenges after disasters. DRR is imperative for sustainable development. Thus, disaster reduction policies and strategies should aim at building disaster-resilient societies with the aim not only to reduce the level of risk in societies but also to ensure that development efforts do not increase the vulnerability to these hazards. It is worthwhile to note here that the capacity of human societies to resist disasters is determined primarily by the internal strength and weakness of the society at the level of social economic and cultural vulnerability. The capacities to cope with the impact of disasters are different depending on social conditions such as poor and rich, men and women, young and old, indigenous and non-indigenous, lower caste and upper caste. In other words, the differential vulnerability to disasters provides the forefront of critical issues related to class, caste, race/ethnicity, gender and inequality.

From a gender perspective, extant literature shows that women are particularly subject to disaster because of environmental degradation, migration, urban displacement and migration, poverty and unpaid caregiving responsibilities (Cutter 1995; Shiva 1988; Steady 1993). Importantly, a gendered division of labour constructs many women to face the brunt of disaster. In sum, it is already discussed in the introductory chapter that gendered vulnerabilities to disasters are the result of structural failures, government's unwillingness to provide adequate funding and training where it was needed, inadequate attention to disaster mitigation, preparedness, response and recovery, lack of maintenance of critical infrastructure, pattern of social inequality within the deprived woman, outright negligence and disregard for human life. In keeping with this emphasis, Enarson and Morrow (1998) noted:

> The social experience of disaster affirms, reflects, disrupts and otherwise engages gendered social relationships, practices, and institutions. Disasters unfold in these highly gendered social systems. Disaster management is correspondingly engendered, shaping the environmental decisions we make and contingencies we fail to plan for the dynamics of our disaster-management organisations and relief operations, the disaster-responding household and emergent response groups, the decision makers we choose and the heroes we create. Disaster is certainly gendered in the discourse we employ to depict and speak about natural disaster, how we theorise and study disaster, who is trained and how, what work is published or funded—and who the 'we' is in disaster practice. (pp. 3–4)

My view is that there is limited long-standing assessment in understanding gender concern within disaster. So, problems will persist as usual time and again unless national and international planning and development programmes as well as academic research integrate questions of gender with effective action at the ground level. Towards that end, I propose few steps in support of gender, development and DRR in the following three phases:

Understanding/Perceiving the Nature of Hazards/Risks

Natural hazards have been occurring since the creation of society and are, therefore, inevitable. Most often, our past experiences on some specific hazards teach us how to understand and anticipate future hazards and consider present situations. Generally, differential perception of risk is found between men and women at the time of preparedness. On the other hand, in all societies and in all types of disasters, women perceive the risks with utmost care. Subsequently, they are more eager to take part in disaster preparedness actions and quicker to pay attention to disaster warnings and evacuation alerts than their male counterparts. Men, who have the tendency to take higher risks, are less likely to respond to evacuation warnings and as a result are more likely to suffer greater losses (Blaikie et al. 1994). As the world learns from each fresh tragedy, however, paradoxically, the omission of potential women from national disaster planning and the dependence on men at large, who are usually less likely to consider early warning messages and less likely to distribute them, can have devastating consequences for men, women and children following disasters. To address this issue, gender-focused research and disaster policy makers are required to explain how and by whom decisions are taken in a community and society in 'normal times' for development purposes, and how does it apply to each of the disaster reduction components? It is apparent that women are essential agents for change and require to be further strengthened. Identifying and marshalling their skills and abilities as social force and channelling it to enhance efforts to protect their communities and dependents is a foremost mission in any disaster reduction strategy. 'Women are also an important target group as risk communicators, family health-care providers, neighbourhood preparedness participants and voluntary community responders after disaster. Researching gendered patterns of disaster communication in diverse cultural, political and economic settings can help practitioners to frame and target information for those most predisposed to listen' (Enarson 1998: 165).

Identifying Vulnerabilities and Capacities

While the gender dimensions of disaster risk impact are complex, it is imperative to understand the diverse axes of social differentiation that separate and connect various women during the cycle of disaster. It is already discussed elsewhere that women die more than men after a major disaster. This is because of social and cultural factors and persistent gender norms, rather than biological ones. Since gender rules differ across countries and cultures, men possibly will be more susceptible due to their risk taking behaviour. Nevertheless, it is also discussed elsewhere that girls and women have limited access to or control over resources, including the possessions necessary to deal with perilous events, such as information, education, health and wealth. Therefore, their vulnerability is comparatively greater than men's. Similarly, psychological responses to disasters include short-term effects such as shock, anxiety, sleep disturbances and guilt. There are differences between women and men and girls and boys in the nature of psychological impact. Voluminous literature has revealed that a greater proportion of women and girls report suffering from emotional disorders and distress as compared to men and boys. There are also instances where women and girls face sexual abuse in disaster situations and as a result were forced into prostitution for basic needs such as food, shelter and security. The sex industry often becomes part of the interaction between the refugee or displaced population and the local community. Men and boys may also be at risk of sexual abuse in such circumstances (WHO 2002).

The fact of the matter is that it is their inequalities in the everyday life, and not just at the time of disaster, that produces greater risk and diminishes life chances for women and girls. Thus, identifying vulnerabilities and capacities is vital to creating lasting change and resilience. Since woman is not a 'homogenous' category, every one of them is not vulnerable all the time and it is not biological difference that defines vulnerability but social constructions of gender roles and relations, and the power inequalities that make the difference of susceptibilities. This reflects the point that reducing gender inequality prior to an event is an important tool for DRR and resilience building. Since women are not a homogenous group, their intersectionality with class, caste religion and age should be thoroughly identified in terms of their access to resources, rights and responsibilities. This will help to understand the social relations that produce and reinforce differences and inequalities within women in any given context with regard to natural disasters. While gender-based social, economic, religious and cultural constructs marginalize women across all communities and groups, irrespective of class, caste, economic standing, status, ethnicity and age, consequently, addressing gender relations in DRR would ensure sustainability of gender balance and empowerment of women with regard to disaster and vulnerability reduction.

One concern is that in many Third World countries, the concept of 'disaster risk reduction' takes a secondary thought. Even at this moment, the old approach of 'disaster management' is used and practiced, emphasizing the engineering and infrastructural interventions in managing disasters. For instance, in many countries, the disaster management planning largely concentrates on land use planning, scientific, technical and professional applications as well as protection of critical facilities, and infrastructure in rural and urban environments. The indigenous and socially mediated efforts including large-scale sociostructural interventions are still missing in action. The point is that with primary operational functions, the approach of vulnerability must be poised to minimize the risk of a disaster. Thus, broadly, there should be a shift towards a 'women in development' perspective, which has an elaborated 'gender and development' approach (Byrne and Baden 1995). Even within the UN system, gender equality, development and peace are incorporated with the hope that importance of gender perspective would help in disaster prevention, mitigation and recovery strategies. Its recommendations included that international and regional organizations must support governments in extending gender-sensitive DRR (UNDAW/ ISDR 2001). Blaikie et al. (1994) rightly point out that risk is originated in conditions of physical, social, economic and environmental vulnerability that require to be measured and dealt with on a regular basis, with the principal idea being to reduce such exposure through the development and reliance on individual capabilities and institutional capacities.

Disaster Risk Assessment and Analysis

A wide-ranging risk appraisal system and analysis, anchored in systematic, methodical and precise information on both hazard and vulnerability, regularly updated and widely disseminated is inevitable for a sufficient and successful disaster reduction strategy (UNISDR 2002). In this regard, the importance to advance the accessibility of gender-sensitive data and its potential to contribute to disaster preparedness and to improve disaster response are discussed.

First, drawing on multidisciplinary gender scholarship, it is acknowledged that women's vulnerability to disaster is deeply rooted in a gendered global economy, so policy makers and concerned gender-focused researchers should see diverse issues by documenting and assessing women's disaster needs in rich as well as poor nations. Accordingly, it is important to identify the gendered division of labour regionally in domestic, waged, agricultural and community sectors; investigate historically and contemporary economic status of women, their work and employment; evaluate the process of feminization of poverty and explicate women's diverse economic recovery resources and income-generating strategies (Enarson 1998).

Second, voluminous literature has shed light on the nexus of gender relations, global development and environmental or technological hazard (Anderson 1994; Anderson and Woodrow 1989; Blaikie et al. 1994; Enarson 1998). It is the responsibility of the disaster researcher and policy makers to document how different women's lives have been made more disaster-prone through globalization, environmental degradation, hyper-urbanization and other massive structural forces more accurately and in broader settings. On the other hand, it is also necessary to address the positive impact of globalization on women's life and their disaster resilience.

Third, disaster planners and policy makers require to acquire intense knowledge on how women and men use urban space and how spatially segregated cities are diversified on the basis of class, caste, gender, race and ethnicity. Scholars have revealed that urbanization is also a highly gendered process (Dandekar 1993; Eichler 1995; Sweetman 1996). There is no doubt that cities provide opportunities to many women, nonetheless, millions of women and children struggle on a daily basis in a hazardous city life (Moser 1996). Thus, strategies require to be widened to include competent women at all stages of policy, planning and management processes. Increasing women's share in political formations, expanding gender awareness and tools for gender competent practice among urban development practitioners, and capacity building among women are believed to be the vital issues. Enarson (1998: 160–161) has directed social scientists and emergency managers to:

- document trends and patterns in women's migration to hazardous urban environments;
- investigate gender and class-specific transport and housing resources in metropolitan regions; and
- contribute to social vulnerability mapping of targeted urban populations, for example, new immigrant women or public housing residents.

Fourth, overwhelming literature has found women's environmental stewardship, their aboriginal knowledge of local resources and their information of family and community life to be an important requirement for degraded environments or environmental crisis. (Agarwal 1997; Women's Feature Service, 1992; Shiva 1988). It is the responsibility of the entire society and government to encourage such brave women so that many more are motivated to take up such work. Instead of looking out women always as victims, when we see them as agents of change, it gives hope for the entire system. Thus, the emergency managers planning for community response to environmental disasters must collect diverse context by which women devote their lives for community disaster preparedness.

Fifth, evidence reveals that family size and structure and power relations in kinship and marriage produce perilous living surroundings for women, particularly in cases of divorce, desertion, widowhood and single mothering (Enarson 1998; Wiest et al., 1994, Winchester 1992). Consequently, the condition of women-headed households has been at higher risk. Moreover, it is also found that gender and kinship relationships structure household food distribution of girls and women after disaster (Rivers 1982). Importantly, study also reveals that domestic violence, sexual abuse and human trafficking make some women at greater risk in the aftermath of disasters (Commission for the Prevention of Violence Against Women 1989). When households lose livelihoods, many young girls and women become easy prey for traffickers. Similarly, women's greater life expectancy rates have exposed many elderly women to face more disadvantages than disabled men (Boylan 1991). Hence, the responsibility of policy makers is to identify and investigate culturally specific day-to-day issues and interlink these with the broad disaster planning. Enarson (1998: 164–165) in this regard has therefore suggested to initiate cross-sectoral research in these areas. Importantly, there must be a clear nexus between academic research and planning initiatives. The area includes the following:

- documenting whether, how and to what extent internal gender relations as well as culturally specific ones in disaster-beset areas affect organizational development, effectiveness and innovation in crisis;
- case studies of factors that facilitate or hinder gender bias and evaluation of change models addressing bias in organized relief and recovery services;
- qualitative and quantitative analysis of women's disaster careers in an era overly concerned with formal credentials, analysing patterns of segregation, integration or resegregation; and
- investigating gender-specific patterns of extra-organizational participation in community mitigation, informal response and community-based recovery initiatives (Enarson 1998: 164–165).

Finally, a successful policy on DRR needs to raise awareness of impending risks among the local population through radio broadcasts, television, local meetings, newspapers and partnerships with community organizations and NGOs (Parida 2008, 2009). Similarly, it is essential to take DRR problems into established platforms addressing poverty reduction, food security, social construction of gendered disaster and development (Parida 2010, 2015a). There is a need for long-term, gender-inclusive engagement to deal with essential vulnerability and poverty factors that influence the resilience of communities to disasters. The United Nations International Strategy for Disaster Reduction (UNISDR 2002) which is a pioneer in DRR at the international level intends to build disaster-resilient communities by promoting increased awareness of the importance of disaster reduction as an integral component of sustainable development. Its objectives are to:

(1) increase public awareness to understand risk, vulnerability and disaster reduction globally;
(2) obtain commitment from public authorities to implement disaster reduction policies and actions;
(3) stimulate interdisciplinary and intersectoral partnerships, including risk reduction networks; and
(4) improve scientific knowledge about disaster reduction (UNISDR 2002).

Moving forward an agenda for gendered DRR also requires the explanations of international agencies. Drawing on WHO (2002), in the context of gendered vulnerability in disaster planning, I merged various policy directions which are given in the Boxes 14.1 and 14.2:

Box 14.1

What research is needed?

- There is urgent need for international data sets to provide sex-disaggregated data on disaster-related mortality, morbidity and long-term health consequences.
- At the local level, there needs to be better documentation of:
  - whether and how perceptions, preparedness and warning response in disaster relief and vulnerability are influenced by gender-based differences between women and men;
  - how gender relations operate in households and communities in a disaster situation and during the relief and recovery phase; and
  - how gender roles and gender relations influence coping mechanisms.

Cont'd...

...Cont'd

- Research is needed both at the local and at the national and regional levels on structural processes and factors that increase disaster vulnerability in women and men across different social groups.
- Domestic and sexual violence in the aftermath of a disaster and the special needs of disadvantaged women, such as women with disabilities and women in violent relationships, needs to be better researched.
- Gender relations in disaster organizations; effects of gendered organizational culture on staffing, funding, programming, training; and the consequences of these for women and men in disaster situations is another important area that needs to be addressed by research studies.
- Evaluation studies of recovery and mitigation programmes are needed to assess effectiveness of different strategies which incorporate gender equality and social justice into disaster mitigation and reconstruction.

Source: World Health Organization (2002).

Box 14.2

What are the implications for disaster relief programmes and policies?

- Pre-disaster activities such as hazard mapping and vulnerability analysis should integrate gender considerations. At a minimum, this would involve taking into account:
  - the differences in vulnerability to and impact of disasters in women and men; and
  - how their roles and status affect disaster relief and recovery.
- Community-based disaster-preparedness projects and disaster training and education programmes should include women and men, and address their respective needs and concerns.
- Information collected through rapid assessments of health status and health needs in disaster situations should be sex-disaggregated and include a gender analysis. Information is needed on reproductive and mental health needs, in addition to information on communicable diseases and on malnutrition.

Cont'd...

...Cont'd

- Basic health services provided as part of emergency relief should likewise cater to women's and men's different needs and be provided in a gender-sensitive manner.
- Disadvantaged groups such as women and men with disabilities and women in violent relationships should be identified as special-risk populations for disaster relief and recovery services. It may be useful to work with community organizations to identify these groups.
- Gender training of emergency managers and health service providers should become an integral part of staff training in all organizations and agencies involved in disaster-relief.
- The initiative for gender mainstreaming the Coordinated Appeals Process (CAP), a tool for United Nations coordination and programming for humanitarian assistance at the international level, needs to consciously address health issues from a gender perspective.

Source: World Health Organization (2002).

Challenges to Gender-inclusive DRR

DRR is a broad approach, which includes all action aiming to reduce disaster risks. Action can be political, technical, social and economic. It takes various issues such as policy guidance, legislation, preparedness plans, agricultural projects, an insurance scheme or even a swimming lesson. This approach enables people to think and work across society, to make sure that everyone—from governments to individuals—makes the right decisions to reduce the risk and impact of disasters (UNISDR 2010). Though the subject of gender in DRR is discussed widely at various levels, the fact of the matter is that there is no progress in mainstreaming gender into DRR. This issue still remains largely marginalized from the DRR process.

Generally, there is a poor understanding of gender aspects in DRR at the policy and practitioner levels, which causes major hurdles in implementing safe development practices, altering beliefs and behaviour, and improving disaster management in order to minimize the numerous and diverse conditions and activities. Gender equality in DRR does not denote purely addressing women's humanitarian issues—it means dealing with the apprehensions of men and women, the relations between them and the root causes of imbalances. In most of the Third World countries, though the concern of gender is included in the policy design at present, it continues to be identified as an 'add on' aspect, rather than an integral component in development and DRR. Indeed, the heavily loaded bureaucratic-oriented planning and policy fails to understand that gender needs to be considered as a central and cross-cutting issue in all aspects of development. The fact is that while women play important roles in building a culture of disaster prevention, particularly at the community level, this is not well recognized and, so, their potential to contribute to DRR is mostly left untapped. Moreover, there is lack of genuine political accountability and financial resources for understanding how women in different life circumstances respond to hazards and participate in relief and reconstruction process. The engineering model of disaster policy is yet to seriously reflect on the issue of the social construction of gendered disaster vulnerability. In the meanwhile, lack of institutional and individual capacity and tools to mainstream gender and DRR is clearly visible. Compared to the issue of development, gender and DRR knowledge and capacity are still acquired by only a relatively small group of professionals and practitioners working in these respective two areas. A majority of the disaster managers and professionals often lack the knowledge required to address gender issues in DRR.

Conclusion

The preceding discussion reflected on gender-focussed planning, policy and the way to future academic research. At the most fundamental level, our society lacks an effective, vibrant and comprehensive disaster reduction policy and as a result women's lives are made more difficult in ways that count in disasters. On the contrary, most women in developed countries are protected by emergency management systems, zoning and construction standards, employment opportunities, good maternal health, high rates of education, antiviolence sheltering systems and other conditions reducing their susceptibility. Consequently, there are low rates of disaster-caused injury and mortality. In Third World countries, women are vulnerable and disadvantaged—socially, economically, politically, psychologically—in their everyday life, therefore are expected to be more vulnerable in disaster situation. My whole argument in this context is that gender-sensitive analysis needs to be incorporated into the study of disasters and emergency planning because development process is so clearly gendered. Since disaster management and emergency planning is 'male-dominated' and 'action oriented' (as discussed elsewhere), a social vulnerable approach would possibly try to fix the problem in the direction of DRR. There is no doubt that understanding historically marginal groups and their extent of vulnerabilities would be step towards meaningful mitigation. On the other hand, social vulnerability to disaster is neither uniform nor universal. Thus, addressing the root cause of differential gender vulnerabilities will result in a sustainable, safer and a more just world. This idea is echoed in several deliberations. It is worth mentioning that we are in a globalized society, so mitigating disaster requires coordination at local, national and global levels. As Enarson (1998) says:

> At the heart of disaster mitigation and risk reduction is the global struggle towards more sustainable patterns of environmental, economic and human development. Because women are central actors in family and community life, gender equality and women's empowerment are, in turn, at the heart of the global project of sustainable development. In this sense, more egalitarian social relations enable societies to pursue more sustainable patterns of growth and development, and hence learn to live more securely with hazard and risk in safer environments. (p. 166)

I strongly support an integrated and holistic approach to risk reduction and focus on disaster and development, community involvement, mitigation and prevention, capacity building and social justice. Practitioners, policy makers, community leaders, emergency managers, activists and scholars concerned about the social construction of gendered disaster have to show potential will power to meet women's immediate needs and long-term interest in disasters cutting across the barriers of class, caste, age, culture and place (Parida 2015a, 2015b). Several chapters in this volume have employed combined physical and social vulnerable analysis. There are a number of methodologies that have been developed through multidisciplinary research that puts forward ways for integration. Noting this as a potential for disaster research and policy, there is immense opportunity for academic researchers, activists and emergency managers. Let's pledge to debate further in support of building more resilient, sustainable and just ways of living in an increasingly hazardous world.

References

Agarwal, B. (1997). "Environmental Action, Gender Equity and Women's Participation", *Development and Change*, Vol. 28, No. 1, pp. 1–44.

Anderson, M. (1994). "Understanding the Disaster-Development Continuum: Gender Analysis Is the Essential Tool", *Focus on Gender*, Vol. 2, No. 1, pp. 7–10.

Anderson, M. and Woodrow, P. (1989). *Rising from the Ashes: Developing Strategies in Times of Disaster*. Boulder: Westview.

Blaikie, P., Cannon, T., Davis, I. and Wisner, B. (1994). *At Risk: Natural Hazards, People's Vulnerability, and Disasters*. London: Routledge.

Boylan, E. (1991). *Women and Disability*. London: Zed Books.

Byrne, B. and Baden, S. (1995). *Gender Emergencies and Humanitarian Assistance*. BRIDGE Report No 35, IDS, Sussex.

Comfort, L., Wisner, B., Cutter, S., Pulwarty, R., Hewitt, K., Oliver-Smith, A., Wiener, J., Fordham, M., Peacock, W. and Krimgold, F. (1999). "Reframing Disaster Policy: The Global Evolution of Vulnerable Communities", *Environmental Hazards*, Vol. 1, No. 1, pp. 39–44.

Commission for the Prevention of Violence against Women (1989). *Violence against Women in the Aftermath of the October 17, 1989, Earthquake: A Report to the Mayor and City Council of the City of Santa Cruz*. Santa Cruz, California.

Cutter, S. (1995). "The Forgotten Casualties: Women, Children, and Environmental Change", *Global Environmental Change: Human and Policy Dimensions*, Vol. 5, No. 3, pp. 181–194.

Dandekar, H. (ed.) (1993). *Shelter, Women and Development: First and Third World Perspectives*. Ann Arbor, MI: George Wahr.

Eichler, M. (1995). *Change of Plans: Towards a Non-Sexist Sustainable City*. Toronto: Garamond Press.

Enarson, E. (1998). "Through Women's Eyes: A Gendered Research Agenda for Disaster Social Science", *Disasters*, Vol. 22, No. 2, pp. 157–173.

Enarson, E. and Morrow, B. H. (eds.) (1998). *The Gendered Terrain of Disaster: Through Women's Eyes*. Westport: Greenwood.

Moser, C. (1996). "Confronting Crisis: A Comparative Study of Household Responses to Poverty and Vulnerability in Four Poor Urban Communities", *Environmental and Sustainable Studies and Monograph Series #8*, World Bank, Washington DC

Parida. P. K. (2008). "Super Cyclone Affected Coastal Orissa: A Social Vulnerability Approach", *Review of Development and Change*, Vol. XIII, No. 2, pp. 159–180.

Parida, P. K. (2009). "Vulnerability of Disaster Victims: A Study from Coastal Orissa", *Eastern Anthropology*, Vol. 62, No. 2, pp. 209–224.

Parida, P. K. (2010). "Understanding Evacuation Behaviour in a Disaster Society: The Case of Coastal Orissa", *Sociological Bulletin*, Vol. 59, No. 2, pp. 179–198.

Parida, P. K. (2015a). "The Social Construction of Gendered Vulnerability to Tsunami Disaster: The Case of Coastal Sri Lanka", *Journal of Social and Economic Development*, Vol. 17, No. 2, pp. 200–222.

Parida, P. K. (2015b). "Natural Disaster and Women's Mental Health", *Social Change*, Vol. 45, No. 2, pp. 256–275.

Rivers, J. (1982). "Women and Children Last: An Essay on Sex Discrimination in Disasters", *Disasters*, Vol. 6, No. 4, pp. 256–267.

Schiebinger, Londa. (1997). "Creating Sustainable Science", *Osiris*, Vol. 12, pp. 201–16.

Shiva, V. (1988). *Staying Alive: Women, Ecology and Development*. London: Zed Books.

Steady, F. C. (ed.) (1993). *Women and Children First: Environment, Poverty, and Sustainable Development*. Rochester: Schenkman Books.

Sweetman, C. (ed.) (1996). *Women and Urban Settlements*. Oxford: Oxfam.

UNDAW/ISDR (United Nations Division for the Advancement of Women and the International Strategy for Disaster Reduction Expert Group Meeting) (2001). *Environmental Management and the Mitigation of Natural Disasters: A Gender Perspective*. E-forum, November.

UNISDR (United Nations International Strategy for Disaster Reduction) (2002). *Gender Mainstreaming in Disaster Reduction*. Geneva: UNISDR.

UNISDR (United Nations International Strategy for Disaster Reduction) (2010). *Disaster Risk Reduction: An Instrument for Achieving the Millennium Development Goals*. Geneva, Switzerland: UNISDR.

Wiest, R., Mocellin, J., and Motsisi, D.T. (1994). *The Needs of Women in Disasters and Emergencies*. Report prepared for the Disaster Management Training Programme of the United Nations Development Programme and the Office of the United Nations Disaster Relief Coordinator. The University of Manitoba Disaster Research Institute, Winnipeg.

Winchester, P. (1992). *Power, Choice and Vulnerability: A Case Study in Disaster Mismanagement in South India*. London: James & James.

Women's Feature Service (1992). *The Power to Change: Women in the Third World Redefine their Environment*. London: Zed Books.

WHO (World Health Organization) (2002). *Gender and Health in Disasters*. Geneva, Switzerland: Department of Gender and Women's Health.

Index

Abraham, B., 177
Agarwal, B., 29
Ahmad, T., 104
Albert, B., Mcbride, R., and Seddon, D., 130
Aldwin, C. M., Levenson, M. R., and Spiro, A., 197
Alway, Joan, Belgrave, Linda Liska, and Smith, Kenneth J., 21
Anderson, Mary B., 21, 32
Angermeyer, M. C. and Kilian, R., 84
Ariyabandu, M. Malalgoda., 56
Ariyabandu, Malalgoda Madhavi and Maithree, Wickramasinghe, 66

Barkun, M., 258
Begum, R., 40
Blaikie, P., Cannon, T., Davis, I., and Wisner, B., 20, 312
Bolin, R. C., Jackson, M., and Crist, A., 41
Bourdieu, Pierre, 56
Britton, N., 35
Brown, W., 234
Butler, L. D., Blasey, C. M., Garlan, R. W., McCaslin, S. E., Azarow, J., Chen, X. H., Desjardins, J. C., DiMiceli, S., Seagraves, D. A., Kramer, H. C., 204

Calhoun, L. G. and Tedeschi, R. G., 194, 200
Centre for Research on the Epidemiology of Disasters (CRED), 272
Centre for Science and Environment (CSE), 31
Challenges to Gender-inclusive DRR, 317
Chaupadi, 159
Chennai floods of 2015, 224
and slums, 225
Childers, C., 41
Chipko Movement, 31
Community Self Reliance Centre (CSRC), 160
Covello, V. T., 184
Crispin, C., Sawa, M., and Taku, K., 204
Cuddalore, 176

Cycle of disaster, 135, 140

De Ridder, D., 204
Deshpande, A., 142
Development Decade for Women, 20
Differential vulnerabilities, 57
Disability, 127
Disasters, 156, 259
Disaster communication, 140
Disaster Management Centre (DMC) of Sri Lanka, 281
Disaster mitigation, 260
Disaster, 19–21, 23, 25, 27, 29, 31, 33, 35, 37, 39, 41, 43, 45, 47, 102, 258
Ditlevsen, D. N. and Elklit, A., 86

Ecofeminism, 29
Emergency response, 39
Enarson, E. and Morrow, B., 34, 57, 59, 309
Enarson, E. and Fordham, M., 22, 34
Enarson, E. and Phillips, B., 29
Enarson, E., 24, 64, 66, 313, 319
Environmental conservation, 239–240

Feminist environmentalism, 29
Feminist theory, 29
Feminization of poverty, 64
Fordham, M., 65
Fothergill, A. and Peek, Lori, 58, 65
Fothergill, A., 33, 58, 65
Fraser, Nancy, 234
Frazier, P. A. and Kaler, M. E., 206
Fritz, C. E., 258

Gee, J. P., 177
Gender, 20–22, 54
Gender and development, 29
Gender and migration, 161
Gender difference, 86, 93–94
Gender differentiation during the process of disaster, 33
Gender in disasters, 53
Gender mainstreaming in disaster, 222
Gender relations and market, 77
Gender relations within community, 75
Gender vulnerability, 104, 120
Gender, dependency and natural disaster, 30
Gender-based disaster studies, major approaches, 26
Gender-based prejudices and divisions, 55
Gendered identities, 56
Ghai, A., 139
Ginige, Kanchana, Amaratunga, Dilanthi, and Haigh, Richard, 27
Glazer, Nona, 32
Gokhale, V., 176
Gordon, L., 189
Gupta, D., 142

Haberli, I., 264
Health, 156
Helgeson, V. S., Reynolds, K. A., and Tomich, P. L., 204, 206
Hozumi, Tomoo, 157

Ikeda, Keiko, 65

Incompetent state, 137
India's disaster diplomacy, 274
India's human assistance programme in Indonesia, 283
India's humanitarian assistance programme,
in Afghanistan, 286
in Bangladesh, 293
in Maldives, 282
in Myanmar, 284
in Sri Lanka, 277
in Pakistan, 290
in Philippines, 289
Indian disaster assistance programme in Nepal, 275
Insufficient social security, 144
International Organization for Migration (IOM), 158, 161

Jenkins, S. P. and Rigg, J. A., 147
Johnsson-Latham, G., 22

Kessler, R. C., Sonnega, A., Bromet, E., Hughes, M., and Nelson, C. B., 86
Krishnaraj, Maithreyi, 66
Kumar, N., 179

Lepcha, 241, 244
Liberal feminism, 67
Lindell, Michael, 258
Linley, P. A. and Joseph, S., 205
Liquor and women's anxiety, 229

Male bias, 20
Marshall, B. K., Steven, P. J., and Schlichtmann, J. R., 258
Matanhelia, P., 190
McEntire, D. A., 27
Media in disaster, 105
Mehta, Manjari, 63, 66
Menon, Nivedita, 56
Mental health and quality of life, 84
Migration, 164
Ministry of Railways (2011) Report, 67
Moghadam, Valentine M., 64
Mohindra, V. and Azhar, S., 180, 189
Morrow, B. H. and Enarson, E., 65
Mothering, 32
Mouzelis, Nicos, 57

Nepal Earthquake, 153, 155, 157, 159, 161, 163, 165, 167, 169, 171, 173, 175
Nirmala, K., 178

Oesterreich, R., 206
Oliver-Smith, A. and Hoffman, S. M., 257
Oliver-Smith, Anthony, 260

Parida, P. K., 34, 40, 58
Park, C. L., Cohen, L. H., and Murch, R. L., 201, 206
Patriarchy, 30
Peacock, W. G. and Ragsdale, A. K., 28
Pederson, E. L. and Vogel, D. L., 205
Peuler, J. N., 37
Phillips, B., 40
Pietruch, M. and Jobson, L., 201
Pincha, Chaman, 65
Polatinsky, S. and Esprey, Y., 206
Post-disaster mortality, 24

Post-traumatic growth, 193, 195, 197, 199, 201, 203, 205, 207, 209, 211, 213, 215, 217
Post-traumatic stress disorder, 83
Poverty/lack of access to resources, 147
Prati, G. and Pietrantoni, L., 202
Preparedness behaviour, 35
Problems handled by women volunteers with recognizing their identity, 187
PTG, 194
  clinical utility, 197
  gender difference, 201
  – age-gender interaction, 202
  – biological factors, 206
  – coping behaviour, 203
  – distress, 202
  – empathy, 205
  – methodological incompetence, 206
  – new social roles, 204
  – possible causes, 202
  – rumination, 203
  – self-disclosure, 204
  – socialization and social support, 204–205
  important predictors, 200
  process of, 198
Purdah, 65

Quarantelli, E. L. and Dynes, R. R., 260
Quarantelli, E. L., 26

Raj, Anitha, 181
Ransom, S., Sheldon, K. M., and Jacobsen, P. B., 206
Reconstruction, 40
Recovery, 40
Rehabilitation and Reconstruction (RR), 263
Reuter, C., Ludwig, T., Kaufhold, M. A. and Pipek, V., 182
Risk, exposure to, 33
Risk perception, 34
Rose, Gillian, 67
Ross, Tom and Lott, Neal, 222

Sathiam, Anita, 184, 187
Schaefer, A. J. and Moos, R. H., 200
Sen, A., 24, 243
Sheldon, K. M., Arndt, J. and Houser-Marko, L., 199
Shiva, V., 29
Shklovski, I., Burke, M., Kiesler, S. and Kraut, R., 140
Simonds, V. M. and Whiffoen, V. E., 87
Singer, T., Seymour, B., O'Doherty, J., Kaube, H., Dolan, R. J., and Frith, C.
Sithara, Sara, 184, 186
Social isolation, 141
Social media, 106–107
Social networking, 181
Social relations framework, 71
Spiegel, D. and Hastings, T. A., 204
Stal, Marc, 105
Stallings, R. A., 260
State and gender relations, 78
Strauss, Anselm and Corbin, Juliet, 69
Sukul, A., 190
Sustainable disaster risk reduction, 307

Taku, K., Cann, A., Tedeschi, R. G. and Calhoun, L. G., 201
Tamil Nadu Floods, 2015, 108
  and social media, 108
Technological support for women volunteers to handle the emergency, 184
Tedeschi, R. G. and Calhoun, L. G., 201
Texeira, R. J. and Maria, G. P., 205
Tharan, Krithika, 182–183
Thoits, P. A., 204
Thompson, S. C., 198
Tolin, D. F. and Foa, E. B., 203
Tomich, P. L. and Helgeson, V. S., 206
Torry, W. L., 260
Traustadottir, R., 130
Treynor, W., Gonzalez, R. and Nolen-Hoeksema, S., 203
2001 Gujarat earthquake in India, 158
2004 Asian earthquake, 36

UN, 20, 30, 312
Flood risk and women volunteers' communication, understanding of, 182
UNESCO, 153
UNISDR (International Strategy for Disaster Reduction 2004), 27
United Nations Children's Fund (UNICEF), 157
United Nations, 162
UNRISD/UNFPA study, 31
Violence against women, 144
Vishnevsky, T., Cann, A., Calhoun, L. G., Tedeschi, R. G., and Demakis, G. J., 203, 206
Vulnerability, 26, 104
  types, 27

Warning communication and response, 35
Wasserman, David, Asch, Adrienne, Blustein, Jeffrey, and Putnam, Daniel, 129
Weiss, H. M., 206
WHO, 315
WHO's Mental Health Atlas 2014, 158
Wiest, R. E., 24
Williams, Stewart, 262
Wisner, B., Blaikie, P., Cannon, T., and Davis, I., 26, 104, 128
Women's communication climate, 180
Women's reliability in relief management with the help of mobile communication, 185
Women's representation and participation in disaster management, 231
World Bank study, 22
World Economic Forum (WEF), 162
World Health Organization, 157, 230
Wortman, C. B., Silver, R. C., and Kessler, R. C., 204

Zoellner, T. and Maercker, A., 197